ANOTHER HAUL

Folklore Studies in a Multicultural World

The Folklore Studies in a Multicultural World series is a collaborative venture of the University of Illinois Press, the University Press of Mississippi, the University of Wisconsin Press, and the American Folklore Society, made possible by a generous grant from the Andrew W. Mellon Foundation. The series emphasizes the interdisciplinary and international nature of current folklore scholarship, documenting connections between communities and their cultural production. Series volumes highlight aspects of folklore studies such as world folk cultures, folk art and music, foodways, dance, African American and ethnic studies, gender and queer studies, and popular culture.

FOLKLORE STUDIES
IN A MULTICULTURAL
WORLD

ANOTHER HAUL

Narrative Stewardship and Cultural Sustainability at the Lewis Family Fishery

Charlie Groth

University Press of Mississippi
Jackson

www.upress.state.ms.us

The University Press of Mississippi is a member
of the Association of University Presses.

Copyright © 2019 by University Press of Mississippi
All rights reserved
Manufactured in the United States

First printing 2019

∞

Library of Congress Cataloging-in-Publication Data

Names: Groth, Charlie, 1965– author.
Title: Another haul : narrative stewardship and cultural sustainability at the
Lewis family fishery / Charlie Groth.
Description: Jackson : University Press of Mississippi, [2019] | Series: Folklore
studies in a multicultural world | Includes bibliographical references and index. |
Identifiers: LCCN 2018032925 (print) | LCCN 2018035224 (ebook) | ISBN
9781496820372 (epub single) | ISBN 9781496820389 (epub institutional) | ISBN
9781496820396 (pdf single) | ISBN 9781496820402 (pdf institutional) | ISBN
9781496820365 (cloth : alk. paper) | ISBN 9781496820853 (pbk. : alk. paper)
Subjects: LCSH: Lewis Fishery. | Fishers—Social life and customs. | Oral
tradition. | Shad fisheries—New Jersey—Lambertville. | Lambertville (N.J.)—
Social life and customs. | LCGFT: Ethnographies.
Classification: LCC HD8039.F66 (ebook) | LCC HD8039.F66 N444 2019 (print) |
DDC 338.3/72745—dc23
LC record available at https://lccn.loc.gov/2018032925

British Library Cataloging-in-Publication Data available

to Adelaide, Keziah, Sarah, and Andrew,
who already know
how to take care of community,
and
in memory of Fred and Nell Lewis,
with gratitude

CONTENTS

Acknowledgments ix
Preface: Through the Gate: Research Methodology and Reflexivity xiii

ONE   Welcome to the Island: The Lewis Fishery in Context 3

TWO   Fishing with Purpose: The Big Stories 29

THREE   The Captains: Between Myth and Legend, Article and Anecdote 57

FOUR   "Were You There When . . . ?": Microlegends 81

FIVE   "It's Like I Said to So-and-So": Everyday Storying 107

SIX   Talking the Walk: Processional Storytelling. 127

SEVEN   Who-All's Coming Down to the Island: Belonging at the Lewis Fishery 147

EIGHT   "A Whole 'Nother Place": Narrative Stewardship and Sense of Place 169

NINE   Fishing in the Mainstream: Anomie, Sustainability, and Narrative Stewardship . . . 193

Appendix A: Map of Lewis Island 217
Appendix B: Lewis Family Tree. 219
Appendix C: Lewis Fishery Catch Statistics 221
Notes. 225
References. 229
Index 239

ACKNOWLEDGMENTS

A twenty-plus-year-old project owes its substance, shape, and fruition to hundreds of people, no lie. Some folks appearing in these pages were not yet born when the project began, and some passed before the project's completion; this research reflects the turning of generations, the essence of tradition. In advance, I apologize to those whose names who do not appear here; acknowledgments sections are always too brief. And, of course, any errors in this work are my own, not the fault of those who have contributed so generously.

My greatest debt goes to the extended Lewis family and current and retired Lewis Fishery crew members, who welcomed me first as a researcher and then as a crew member and family friend. The names of many of them appear as participants and interviewees in the pages that follow, but the help of the late Fred Lewis, the late Nell Lewis, Muriel Lewis Meserve, Steve Meserve, Sue Meserve, and Pam Meserve Baker requires special mention. Without their knowledge, patience with my questions, thoughtful responses to my interpretations, sharing of data and photographs, reviewing of drafts, and many other kindnesses, the work would be a big mess. Several photographers have been particularly generous sharing their work with me during this research: Adelaide Groth-Tuft, Randy Carone, Frank Jacobs, Mary Iuvone, Robyn Stein, Paul Savage, Stephen Harris, and Lita Sands. Visitors to Lewis Island, including fishery customers and local residents, have continuously built my understanding of what the place means and how it works.

This project began my second semester in the Folklore and Folklife program at the University of Pennsylvania, and professors Margaret Mills, the late Roger Abrahams, Robert Blair St. George, and Regina Bendix shaped this work while training me. In the public sector, mentors Rita Moonsammy, Betty Belanus, and Gail Stern enriched this work with research opportunities, feedback, and encouragement. Audience members at American Folklore Society annual meetings contributed through their questions, comments, and interest. Members of the 2016 Folklore Studies in a Multicultural World workshop, especially Craig Gill and Simon Bronner, guided my revisions just when and how I needed it. From the project's inception, Debra Lattanzi Shutika affected this work profoundly with sound advice and ever-present faith that I would produce something worth reading.

My colleagues at Bucks County Community College and the New Jersey and Pennsylvania campuses of DeVry University have supported me with their scholarly feedback and good humor about my fishing schedule. Invaluable research assistance came from the late Yvonne Warren of the Lambertville Historical Society, as well as the staffs of the Lambertville Free Public Library, Bucks County Community College Library, New Jersey State Library, Hunterdon County Library, Newark Public Library, and Mercer Museum Library. During class, my students at Bucks, DeVry, Penn, and Rider University kindly listened to numerous examples from my research; having them in mind tempered my prose, preventing it from becoming a jargon mountain. At the University Press of Mississippi, Craig Gill persistently encouraged me over years, then Katie Keene and Mary Heath ably took up the project at the stage of finally, actually becoming a book. I thank all the staff at UPM, who patiently tolerated my many questions and reassured me that it was OK to ask. As the copyeditor, Norman Ware not only applied great skill but also made the process pleasant through his collegial approach. To the anonymous reviewers, thank you for your thoughtful comments, which undoubtedly improved the manuscript and moved the process along.

And then there are the family and friends "in my corner." Generous listeners have faithfully followed the project over two decades, including Julie Cullen, Liz Lacey, the late Leanne Parks, Barbara Hamilton, Sharon Hallanan, Rory Osler, and the girls and coadvisers of my Girl Scout troops and the Religion in Life Alumnae group. My parents, Bill and the late Pat Groth, provided babysitting in the early years, and my siblings the Reverend John Groth, the Reverend Dr. Bobbie Groth, and Dan Groth helped with elder care in the later years. Bess "Why-not?-It's-something-to-do" Rumball (my godmother) and curious Cousin Ruth went along on fishing hauls when most septua- and octogenarians would not. My sister and eldest brother, Bobbie Groth and Bill Groth, caught on early to the magic of Lewis Island; their enthusiastic expectation of a book to read was more motivation than they realize.

This project probably would not have lasted past spring 1996 without the commitment of my immediate family: my daughters Keziah and Addie Groth-Tuft, and my husband Daniel Tuft. Baby Keziah attended my first interview, and Adelaide was born into fishery life. Later, Adelaide's photographs supported many conference papers, and Keziah contributed detailed manuscript feedback from an undergraduate's point of view. At first, my daughters did not have any choice about participating, so with much admiration I acknowledge that their devotion to the Lewis Island community has grown through their own volition, costing them time, sweat, homework on the riverbank, and many, many late dinners.

My husband Dan has shown great enthusiasm and carved out uncountable hours at home for my work, which he allowed to become his own work when he joined the crew a couple seasons after the research started. Whenever Captain Steve asks, "Shall we do another haul?" Dan reliably responds with a cheerful "Sure!" I suspect that Steve directs this question to Dan *because* his answer is so reliable. And then the crew loads up the net again, more stories are told on the riverbank, and the tradition continues.

PREFACE

Through the Gate: Research Methodology and Reflexivity[1]

On the point of Lewis Island in Lambertville, New Jersey, the air mixes with the smell of the nets still wet with river and shad slime, the same slime that sticks fish scales to my daughters' faces. It's also the slime infused in the knitted backs of my work gloves. They stink out loud, so loud I think the gloves would walk away, stiffened with stink, if they weren't pinned to a clothesline up in the crew's cabin. When I come over the bridge and through the gate ready to fish each night, my gloves are often still damp and cold from the previous evening. I have to put away my mainland self, have to take a leap of faith and put my hands in those horrid gloves, commit myself to being on this island and nowhere else. Commitment isn't always pretty . . .

I never expected to spend my springtimes fishing. My brothers used to hate to take me line fishing with them, because I talked too much, so it is little wonder that I am involved in a fishing operation where talking doesn't interfere with fishing. In fact, talking while fishing is the focus of my study. "Talking at the fishery" refers to various types of storytelling, or narrative, in particular, but also the verbal context, ritualistic behavior, and worldview that surround what people say and experience on Lewis Island. In short, I bring a folklorist's lens to cultural expressions and structures sometimes difficult to pinpoint, in part because they reflect the ethnic mainstream, an unmarked category. In the opening of her book about Fort Wayne, Indiana, flood stories, Barbara Johnstone declares, "I intend this study to help fill a gap in sociolinguistic research about Americans by focusing on a relatively unstudied population: the nonminority 'mainstream' of the American heartland" (1990, 3). I intend to do the same thing for a different region, and the northwestern quadrant of New Jersey appears to be a particularly undernoticed part of the state (Groth 1998). The fact that the region is primarily made up of the ethnic mainstream also pushes it under the ethnographic radar. Johnstone makes the astute observation that "researchers often treat the 'mainstream' as a monotone background against which populations are foregrounded" (1990, 4)—and this may be particularly the case when looking at traditional, vernacular, or everyday cultural expressions not categorized as "art," which can be more easily foregrounded, presented, and analyzed.

When I began studying the Lewis Fishery in Lambertville, focusing on an embedded cultural practice helped me solve a personal practical problem: support my family and care for my infant while going to graduate school full-time, over an hour away from home. Choosing a field site in my own town for my fieldwork theory course helped make the project of school, home, and work actually happen. I was aware, of course, that I needed to get out of my own direct experience to better learn *how* to research, and so I chose the tradition of haul seine shad fishing, something I had never seen; I only knew that it was central to the town's Shad Festival and that my former neighbor was on the fishery crew.

The specific natures of my topic and my biography led to a pronounced blending of various roles that then significantly affected the ethnography. I had lived in town for about five years when the fishery study started, and it's the kind of town where you have to have lived *before* "back in the day" to be considered "from" there. I was born and raised in the bordering school district, even close enough to ride my bicycle to Lambertville in my teen years. However, my insider status wasn't clear or solid in the town or at the fishery. The spring after that first fieldwork project, my husband and I began coming down to the fishery each night to watch the haul like other Lambertvillians. Over the next couple of years, we became friends with various members of the fishery family, including husband and wife Steve and Sue Meserve, who were in the process of inheriting the fishery leadership roles from Steve's grandparents, Fred and Nell Lewis. When a position on the crew opened with a retirement, my husband was recruited. I continued to observe as a grad student and filled roles around the fishery like other women do: hauling nets, raking the island point, recording data, and helping sell fish. After another retirement, I was recruited to officially be the "assistant" woman, helping Sue Meserve manage sales and data collection on a daily basis under the crew's cabin. In another year or two, I was the second woman ever to be recognized as a full member of the crew, earning a full share. Now I have performed every task men on the crew typically do—except serve as captain, of course—and regularly do work typically done by the women. My participant observation research has been comprehensive as well as continuous through two decades.

While I became a fishery crew member, other related roles developed. As the folklorist for the Northwest Jersey Folklife Project, my public sector duties included work with this fishery. As a resident of the town, I forged close friendships with the extended Lewis family through *town* roles: being a parent in the public school where Muriel Meserve (Fred Lewis's daughter and

Steve Meserve's mother) was a teacher, and her daughter, Pam Baker (Steve's sister), was a parent. Pam and I have collaborated on Girl Scout activities with our daughters. I'm the point person when my church works with their church at Lambertville's Community Kitchen, which Muriel heads. Muriel has looked after our younger daughter after school, and neither of my girls has ever known a time when the fishery family was not central to the "village" that "raised" them, as the proverb goes. Muriel has stood in as "grandma" at my children's school functions, and my husband and I served as pallbearers for her father. Then, in a surprising twist after about a dozen years of research, we learned that Steve, Pam, and I are actually blood relatives on the Meserve (not Lewis) side. It's a distance of about sixth cousins, to be sure, but nevertheless it complicates the situation when reporters or fishery visitors ask me, "Are you part of the family?"

When introducing the concepts of "emic" (from within the culture under study) and "etic" (from outside the culture under study) in cultural anthropology classes I teach, I offer my students my position in the field as a case study: outside researcher, *in* town but *from* the next town, wife of crew member, crew member, mother of crew members, one of the longest-running crew members—and still a researcher. We trace my growing interrelationship with the town and note that I have lifelong relationships with island visitors and townspeople independent of the fishery and my studies. We consider that I share the regional identity (Middle Atlantic) and ethnic grouping (Anglo–northern European–American) with most of the fishery crew. As a crew member, I am an outsider by gender in a male-dominated occupation, but this crew has continually included more and more women and girls since the 1970s. As if running along the surface of a Möbius strip, I, like some other researchers, tread both sides of familiar dichotomies: emic and etic, academic and public sector, public and private, subject and object.

Discrete roles are even more of a luxury for ethnographers today than in the past. Embracing a hybrid researcher identity is necessary for making some folkloristic work possible, and this approach may also provide ways to organically understand those cultures and communities we study and serve, maintain some cohesive sense of self, and stay relevant to the audiences we wish to reach. Having started in 1996, I have spent over two decades performing participant observation even in years when my job duties prevented any other scholarly activity. I myself have gone from young adulthood to middle age with others on the crew, have attended funerals of crew members and their parents, and have watched fishery kids go from babyhood to adulthood, or teenage to parenthood. Having so long to study a single site day in and day out is a rare opportunity, affording one both the advantages and disadvantages of emic and etic views, begging for a reflexive review, that is, a review of how one's social positioning affects the study.

Folklorists as well as sociologists and anthropologists (Alleyne 2002, 610) have a rather long tradition of separation between researchers and the communities they study, reinforcing the illusion that distance creates the objectivity required of social science. We also have a tradition of distaste for "going native" (Rosenberg 1996, 145) and fear of "becoming too close to one's subject." Yet, since the advent of feminist ethnography, postmodernism, and history of science, we have become aware that the social sciences—as well as the physical and natural sciences—do not neatly realize the austere myth of objectivity, for both proximity *and* distance can create biases and blind spots. Our inclusion of "Western-trained" ethnographers studying their own "non-Western" cultures poses more questions about insider-outsider status, as well as leading to the possibility of Western ethnographers studying their own Western cultures. In graduate school at the University of Pennsylvania, we learned to acknowledge our positions vis-à-vis the community under study, and to allow the reader to judge whether position is a help, hindrance, or mixed bag.

The greatest downside of emic study is that it is perennially challenging to study one's own culture regardless of which culture that is (Evers and Toelken 2001; Rosenberg 1996; Stone 1998). It has become common knowledge that one misses things because of assumptions or a "can't-see-the-forest-for-the-trees" situation (Johnstone 1990). For example, I noticed as I transcribed interviews that my interviewees often use the phrase "you know" liberally and sometimes literally, making references to our common experience or information they expected I knew already. They were often right about my knowledge, but I nevertheless had to carefully ask for clarification "for the recording," or examine each quotation's context to double-check my understanding.

And yet, as Ray Cashman (2011) observes about his fieldwork, as one gets to know more about the community, there are more things one cannot share in the public discourse, scholarly or otherwise. One conversation on the riverbank with a film student crystallized for me the difference between our disciplinary ethics: his greatest obligation was to his audience to give them the story, while my ethical framework as a folklorist, as a social scientist, emphasizes instead the obligation to do no harm to the community I study. With time, one comes to understand what counts as culturally sensitive information that should not be treated publicly, and as trust builds, one learns information that, if shared with a general audience, would make informants uncomfortable. In contrast with classic subjects of anthropological study generations ago, this quandary becomes even more significant when one studies people who are highly literate, mobile, and well connected, in one's own geographic area. One's subject is always one's audience, too, and thus the ethnography is socially embedded.

To address the downsides of my emic positioning, I will at times pause in the argument to highlight some personal detail that might impact the interpretation, or to analyze the potential for bias. At the outset, I will overtly state that Keziah and Adelaide Groth-Tuft are my daughters, Keziah being the infant in my first season of study and Adelaide born five years later. Crew member Dan Tuft is my husband. As I move ahead in the chapters, I will usually just identify Adelaide, Keziah, and Dan as crew members, not family, as the relationship is not of primary importance, and since readers already know which people make up my family, they can judge for themselves potential bias.

In the interest of frank and clear disclosure, I must also call attention to the fact that I share the ethics of the group and material I study: I highly value the local environment, local community, local traditions, kindness, and civility. However, my goal is not to convince my readership to value the same things, but rather to analyze *how* the culture supporting these ideals operates in this place so that (1) this research site and subject are documented, (2) we understand cultural systems better, and (3) those who do share the ethics of this place can apply its strategies to other places. The fact that I have raised two of my informants as well as participated fully as a crew member clearly points to the potential for influence on the material I study. Because the values of the fishery culture matched the values of the family culture, I did not "experiment" or remove my children from their own culture while we were all in the field. Instead, the situation bears some resemblance to action research in the scholarship of teaching and learning: the professor teaches a completely legitimate class and the students learn what they need to learn, but along the way the professor-as-researcher collects data, makes observations, and draws conclusions. That said, I share here that whenever I took on the at-this-moment-I'm-a-researcher-not-your-mother stance while interviewing or reviewing collected data, I saw things through my daughters' eyes that I did not see on my own, just as with any other informant. For example, Keziah's astute observations from a teen's perspective about the changing nature of the captain role surprised me and informed my interpretation. Likewise, when I asked seven-year-old Adelaide to photograph people talking while they got ready for the haul because I was busy mending net, I could shift from first to third person to literally see myself in my normal crew role without disrupting the work. I could also view other photos she took of the island as evidence of her sense of microplaces within the space (e.g., the cabin stairs, the bench, the gate).

Besides the efficiency and other benefits of action research, examining my own mainstream culture and community encourages collaboration with subjects (see Lawless 1993 for a full discussion of "reciprocal ethnography"). When we say that insider researchers "know the culture better," in part we are saying that local fieldworkers have a better grasp of what American studies scholar Kent Ryden calls the "invisible landscape": "For those who have developed a sense of place, . . . it is as though there is an unseen layer of usage, memory, and significance—an invisible landscape, if you will, of imaginative landmarks—superimposed upon the geographic surface and the two-dimensional map" (1993, 40). The local researcher then may have a full view of an invisible landscape, and when sharing my experience of "invisible landscape" as a local researcher, I have followed anthropologist Edith Turner's insight that "experience" is "a human birthright" that researchers can and may access (1992, xiii). For example, when the crew looks across Island Creek to the mainland, we can all see the mixed building materials that make up a set of condominiums. Those of us from the area can see the restaurant that burned down in the late 1970s (the stone parts remain), but only I can see my mother, grandmother, and I at my grandmother's birthday lunch there when I was about thirteen, just as others from the area see their experiences in that restaurant as part of the invisible landscape. With my own sense of invisible landscape from lifelong experiences, I already have material with which to participate in conversations about the place, enhancing the *participant* part of participant observation. The invisible landscape is an integral part of the place observed *and* of the observer, even more so for the local fieldworker, who works with no expectation of going home to somewhere else at the end of the project. Local researchers shift between homeplace and research-place, living within both realities at once, seeing the invisible landscapes of researcher, informant, and potential audiences.

Thus, a local researcher can enhance the nature of the fieldwork and potentially yield greater depth and nuance in the resulting analysis. Over the years, the people I study have asked about my findings, and I have shared those findings in all their tentativeness in casual conversations, as well as directly asking for responses to ideas. As I reinterviewed, I found that informants agreed with, tweaked, or furthered my findings, which was immensely reassuring. When I went back to transcribe and/or reread early interviews, I had an odd sensation of déjà vu: transcribed text sounded like it had been created to support points formed later on. In reality, my conclusions had grown organically from collaboration with my informants—both in fishing and in interpreting what the fishery and the island are "about." Muriel Meserve, one of the fishery's owners, read a full draft of this work and had the opportunity to give feedback. Mainly, her comments helped fine-tune the accuracy. However, she also commented that as a social studies teacher, she was aware of many of the concepts but had not thought of applying them to her own family. She felt that the interpretation worked and found it very interesting. (Thank goodness.) Following the excellent suggestion of one of this manuscript's anonymous reviewers, I later asked Muriel whether knowledge of my academic findings had in any way altered how she told stories or what she thought about her narrative activities, and emphasized there was no "right" answer to the question. She responded that her storytelling practices had not changed, but she had become more aware of what she did and how she did it. Muriel Meserve is always both a practical doer and an abstract thinker, and here she became more aware that she was *experiencing* and *contributing to* cultural processes while continuing to do her regular conversational work.

Theoretical Contexts

My own early attempts at interpretation fell flat, probably because the richness of the material made some conclusions obvious, while more nuanced interpretations were beyond my knowledge base at the time. Listening to the reverence with which so many people talk about Lewis Island, I realized that spirituality would be a key area for exploration, and concepts such as stewardship, ritual, faith, and myth have become useful. On joining the Department of Language and Literature at Bucks County Community College in 2009, my thinking took a fruitful turn. When preparing to share my research with my colleagues, many of whom have literature degrees (as do I), I tried putting the verbal arts at the center, and the idea of narrative stewardship and a narrative typology quickly emerged.

Sociologists David R. Maines and Jeffrey C. Bridger succinctly declare, "Communities cannot exist without stories" (1992, para. 10), and folklorists have long delved into the technical and functional differences between story types, and into storytelling processes within cultural contexts. This study examines "situational" storytelling as opposed to "professional" storytelling; narratives are told in the course of both social interaction and "conscious cultural" storytelling, wherein the teller is aware of passing on cultural information (see folklorist William Wilson's schema, discussed in Stone 1998, 26). The concept of "situational storytelling" overlaps with what communication scholars Elisia L. Cohen and colleagues call "communication infrastructure," which they define in part as "a storytelling system . . . that affords residents the opportunity to reflect on and to tell stories about their daily lives" (2002, 221–22). The subject matter also enacts folklorist Linda Dégh's concept of "multi-conduit transmission" (1995), in which stories in various genres are shared from person to person over time and space with various tellers and listeners who have among them specializations, overlapping material, and gaps in material—and thus texts crossing paths with other texts. Moreover, given that this study explores mainstream American culture, I join language education specialist Viv Edwards and classicist Thomas J. Sienkewicz in entreating readers to "think in terms of an oral-literate continuum which recognizes the skills of the oral performer and, at the same time, does not view the literate person as hopelessly cut off from oral skill.... [T]he presence of literacy does not remove all trace of orality, nor must an oral culture always function independently of literacy" (1990, 6). The same can be said for media literacy, and throughout this study readers will find that the people under study intermix oral narration with reading, viewing, and commenting on a variety of written and digital sources.

Early on in the development of this project, the concept of "place" became significant, reflecting on scholarly trends in seeing places not as static sites but as dynamic concepts that people build, negotiate, experience, and use. This study reflects all four "categories or layers" of local lore identified by Kent Ryden: "the depth and intricacy of local knowledge of the natural and physical properties and limitations of the geographical milieu" (1993, 62–63); "the intimate and otherwise unrecorded history of a place" (63); the provision of "a strong sense of personal and group identity" (64); and "the emotions which local residents attach to their place and the components of their place" (66). Sifting through these layers, I hope to show how invisible landscape comes to exist, specifically identifying the process and structures—sensible or imagined—that entwine people, place, and cultural expression, creating place from space (Tuan 1977).

Finally, I also want to make visible and understand traditional, community culture as one site in mainstream America at a point of rapid social and technological change. Immigrant traditional cultures are often studied to see why and how they "survive," change, or "die," identifying factors such as assimilation, economics, and changes in social structures and practical circumstances and often seeing the immigrant cultures as the past. Mainstream American traditions are also often prematurely assumed to have died, but with a twist: they are thought to change not by moving through space but rather by moving through time. Mainstream traditions are assumed to die away as an inevitable function of time's passage—especially, as the common theme goes, when encountering "progress," a generations-old, techno-economic, American mythical ideal. This work considers aspects of local American culture that are often assumed to be insignificant until they are assumed to be no longer in existence. The approach to studying culture, then, considers culture to be a dynamic process to which change and sustainability are both relevant.

To those readers who want "the skinny" on this book, I offer the following summary of my argument. There is a system of narrative at Lewis Island that combines verbal activities, everyday activities, and ritualistic activities in a way that creates and disseminates stories *through* both telling and doing. This narrative system supports caring for the environment, traditions, and community, forming a phenomenon I call "narrative stewardship." Tellers and audience members perform narrative stewardship through various story types and channels. Narrative stewardship is part of an ethical system that is rooted in place and that emphasizes community (sharing, caring, and connection) and commitment (hard work, faithfulness, and tradition). The embeddedness and intricacy of this system makes it invisible, but careful study can help us unpack and untangle what people do, often unconsciously, revealing the power of invisibility to foster sustainability.

In what cultural anthropologists call a "thick description" (Geertz 1973c), the first chapter details the basic elements of the book's cultural context, introducing readers to traditional haul seine fishing, shad, Lambertville, Lewis Island, and the extended Lewis family. The ensuing five chapters (2–6) will each concentrate on a particular narrative genre or phenomenon that together create the system of "narrative stewardship." Chapters 7 and 8 use the typology to discuss belonging and sense of place. The final chapter turns outward, asking readers to consider cultural sustainability and change. That chapter examines the power of committing to narrative stewardship as a cultural tool, thus creating an embedded and organic agency, and pointing to a possible application of the study's findings. Throughout each chapter, readers will find my efforts to interweave data and theory, description and interpretation, in an attempt to follow the model Henry Glassie describes in *Passing the Time in Ballymenone*: "[I]f work is good, old categories will slip and shift, and then melt away as we find the place where social science joins the humanities, where art and culture and history, time and space, connect, where theoretical and empirical studies fuse" (1982, xiv). Thus, the writing voice will shift from describing scenes, to narrating experiences, to presenting quotations that impart the narratives and viewpoints of the people under study, to explaining and applying scholarly theory, to making points of my own. Hopefully, the resulting mosaic will enable readers to comprehend the complexity and richness of this cultural material that I have come to understand across a generation of fishing seasons, one haul at a time.

ANOTHER HAUL

ONE

Welcome to the Island: The Lewis Fishery in Context

When you're on Lewis Island, you have that feeling of being on top of something, something you need to hang onto with your feet. Maybe it's because the bridge bounces a little, and maybe it's because you are on an island with visible, moving water on three sides at the southern point. Maybe it's because you can still see signs of the last floods—maple branches, leggy grass, and plastic grocery bags hanging at the same level eight feet up in scrubby trees, dusty dry but still pointing the way downstream. It's a place where you become very conscious of your ankles, because the riverbank says, "Mind your step." Rocks will roll, and while they tumble toward water, you may too—or worse: you may catch on an exposed root or a tuft of green, brown, tenacious grass—the kind that whistles with new green but shows last winter just below the verdure, the kind that can flip you arse over teakettle. All along the outside path, you watch your step and remember the last time you slipped on spring mud, only your screeching ankles holding you in line.

On the inside path, however, green crowds in to greet you, or perhaps to hide you. You can hear the steam whistle of the train on the other side of the river in Pennsylvania, but you are in a different state. You are where the canopy pushes down and demands that you look for the white violets among the purple ones. You look for the spring beauties early on, their pale pink stripes the color of babies' tongues, but faint like painted china. You look for the Dutchmen's breeches, wondering whether Dirk just left them, him run off into a Washington Irving story. You look for May apples, first tentative like half-closed umbrellas, then claiming the floor by mid-May. You are just a visitor, although you feel you've always been there, the tree roots reaching up to grab your soul.

People often do not notice Lewis Island as they cross the Delaware River between New Hope, Pennsylvania, and Lambertville, New Jersey, by car, but if they amble across the footbridge from Lambertville's mainland to the island, they are in "a whole 'nother world," a world whose stability belies the underlying tenuousness of water and earth. Rope ties together the bridge decking as a way of saving the lumber the next time the bridge washes out—and it certainly will later if not sooner. The boards give a little. Summer dust shifts with each step as water hangs in the air. Pushing through the raggedy gate and avoiding the assemblage of poison ivy and Virginia creeper that flank it, the visitor is at the Lewis Fishery, the only traditional haul seine fishery left on the nontidal Delaware River.

Before they start imagining scenes from saltwater fisheries, readers should have at least the barest outline of traditional haul seine fishing, done at this fishery with a rowboat and 50 to 250 or more yards of net. While one end of the net is walked down the island bank, the other end is unloaded from the back of the boat as it is rowed out into the river then turned downstream and back to shore, landing downstream of the other end to make a *C* shape when viewed from above (see the map of Lewis Island in appendix A and the photographs throughout the book for images of typical hauls). Both ends are pulled to shore by hand, capturing the fish. No rods and reels. No pulling the net into the boat. No "setting" the net or letting it drift for any considerable length of time, just one continuous motion, which constitutes a "haul." Using this traditional haul seine (net-pulling) method, the Lewis Fishery operates nearly every day throughout the spring.

Bill Sr. started fishing at this spot in 1888, and members of his family have been doing it every year since, without exception. Perhaps more impressive than longevity, though, is the *nature* of the persistence through five generations and four captains: William Lewis Sr.; his sons, Fred and Bill Jr.; and then his great-grandson, Steve Meserve. The Lewis Fishery crew even fished about thirty hauls in about as many days in 1953 and 1956, years when the shad population had sunk so low they did not net a single shad all season. Blip. Nothing. That's quite a fat goose egg for a busy six-man crew to pursue for so many weeks, but they did. This same dedication to the shad prompted William Lewis Sr.'s legendary badgering of state, regional, and federal authorities, which then led to the cleanup of the river and the return of the shad. The shad population's health still concerns fishers and biologists, although the river's cleanliness no longer poses the problem. It's an ongoing story of fish that rests upon a body of stories about tradition, the environment, and the community. The Lewis Fishery is a place that creates stories, and stories, in turn, create the place. Visitors feel almost immediately that this place is special, but the "why" is elusive; it's taken me two decades to figure out why, how, and to what end the unassuming cultural magic works here, and that is the subject of this book.

Where in the World: Lambertville as Our Starting Place

Scholar Kent Ryden writes that "the depth that characterizes a place is human as well as physical and sensory, a thick layer of history, memory, association, and attachment that build up in a location as a result of our experiences in it" (1993, 38), something he calls the "invisible landscape" (1993, 40). The "our" here refers to "us" as individuals with individual experiences and perceptions, and also to "us" as a collective group with a collective experience or heritage. Moreover, just as individuals' actions add to the "thickening" of the place where those experiences happened (Casey 1993, 253), when individuals *learn* what has happened in the place when they were not there, the place "thickens" further. Anthropologist Clifford Geertz portrays culture as a "context, something within which [elements such as social behaviors and institutions] can be intelligibly—that is, thickly—described" (1973c, 14). Thus, the thickening of a place is its enculturation, and for this reason, this chapter starts with what folklorists and anthropologists call a "thick description."

Lewis Island is a mile-long island, about seventy feet wide at its widest point. Its geographic and social context adds the depth of heritage, and its history as a family home and business adds another layer. Originally hunting and fishing grounds of the Leni Lenape Chief Copponnokous (Petrie [1949] 1970, 16), the island then became part of the Holcombe Grant from the British king (Gallagher 1903, 6). According to a sign erected by Art Lupine, then employed by the agency now called the New Jersey Division of Fish and Wildlife, Richard Holcombe started operating a shad fishery there in 1771. Still officially named Holcombe Island, this place is more often called Lewis Island after the family that has owned the island's southern half since 1918. Divided from Lambertville's mainland by Island Creek, and from New Hope, Pennsylvania, by about a thousand feet of Delaware River, Lewis Island is private property. The completely undeveloped northern end has an absentee owner. The southern end of the island, today owned by sisters Muriel Lewis Meserve and Sue Lewis Garczynski, has a landscape that suits its family, community, and haul seine fishery purposes.

Ever since an early twenty-first-century flood swept away a shed, leaving behind the heavy hardware stored in it, Lewis Island has had only two buildings. Most prominent is the three-bedroom home with two stories lifted above a ground-level basement, ready to house floodwaters if necessary. Fred Lewis, one of the home's last two steady occupants, recalls:

> In 1933, after I got out of high school... it was in the Depression and my brother and I had fished that season... (we got finished with the fishing around the tenth

of June) and Dad says, "Well, you boys don't have a job; you may as well build a house." So, we started building the house on the island then. And by October we could move in.... Then ... I lived there until we got married. (March 1, 1996)

Indeed, Fred rowed his new bride across the creek on their wedding night, because the bridge was once again washed out by floodwaters, a recurring nuisance for the rest of their sixty-eight years of marriage. Down the island from their home lies the fishery's building, often called a "shack" by outsiders, but the family and fishing crew refer to it as "the crew's cabin." Here, the fishery crew gathers upstairs for warmth and boot-changing and the women sell fish "under the cabin," in a space outsiders think of as the bottom floor of a two-floor building.

The island holds a few other human-built structures, including the one nearly every visitor touches: the footbridge across to the island. Theodore Lewis, brother of the first Lewis owner of the island, paved the tip of the island so the crew could pull the net over stones and cement to keep the fish clean of river mud. A partially stone-paved path, also built by "Uncle Dory," has supported the fishing crew's daily march up the riverbank for near on a century. A handful of park benches set back from the paths give visitors a view of the river and the activity on it. Between the path and the water's edge, there are two sets of cement stairs called "the steps by the house" and "[the steps to] Jake's dock," places that the captain refers to when telling crew members where to "catch" the boat and drag in the net's "sea end" during a haul. Jake and Jake's dock clearly call up an image for the captain when he makes the command, but many crew members know only the crumbling steps, evidence of a story and a trusted relationship. The last sign of civilization is the "center path," which leads from the footbridge at the island's point to an opening in the privet hedge that surrounds the house, then through the wildflowers and trees, up to a garden. In the early 1980s, a neighbor and native of Switzerland, Kathy Berg, asked Fred and Nell Lewis if she could clear a little spot halfway up their island property to establish a garden in the tradition of her Swiss home. The garden intermixes flowers, fruits, and vegetables—large poppies next to tomatoes and gooseberry bushes in raised beds with paths in between. The area's many intrusive and ravenous deer must stay outside the high fence, but two-footed visitors can duck under the arbor, turn the latch, and enter this idyllic spot with a small Swiss flag announcing a tradition of political neutrality if not inner peace. Near the garden, two branches of the path lead to a break in the trees on the riverbank. The rest of the island is ceded to native trees, plants, and other wild things, including some invasive plants and the region's ever-present deer and groundhogs. Someday, the whole island may become completely wild, for Fred Lewis entered into an agreement with a regional conservation organization in an effort to ensure that the island would never fall into the hands of developers.

The river by Lambertville is relatively shallow. In summer, people can wade almost halfway across and still be only waist-high in water. A deeper channel midway is treacherous to waders but home to catfish. Lewis Island lies about fifteen miles north of a significantly shallow section of the Delaware River sometimes called the Little Gap, which is marked by Scudder's Falls and Trenton Falls, where cars cross the river at the I-295 (formerly I-95) and Route 1 bridges, the major north-south arteries that connect New York City and Philadelphia. The Little Gap divides the tidal Delaware from the nontidal Delaware, a distinction that not only defines the Lewis Fishery as the only traditional haul seine fishery on the nontidal Delaware but also prevents ships from traveling past Trenton, a detail that shaped the economy and culture of communities north. Before it was undermined by the railroad system in the late nineteenth century, the Delaware Raritan Canal was the major conduit for goods between Pennsylvania and New York (Delaware and Raritan Canal Commission 2012). Except for several statues in the area and a tourist ride in New Hope, the mules that once towed barges along the canal are gone today. Nevertheless, the towpaths, locks, and many historic structures are on the National Register of Historic Places, and together they form one of New Jersey's most popular state parks.

North of the Scudder's Falls Bridge, one finds what I call "bridge communities" (Groth 1998, 25–26), communities that span a bridge crossing the Delaware and relate more to each other than to the rest of the counties and states they lie within. The Lambertville–New Hope area is one such bridge community, which, although in two different states, shares an ambulance squad (in Lambertville) and a grocery store (in New Hope). The two communities are so tightly associated that occasionally tourists in Lambertville will say that they are not local but from New Jersey, not realizing they are still standing in New Jersey. With a combined population of about 6,500, Lambertville and New Hope are not suburbs associated with New York and Philadelphia, respectively, but rather are small towns associated with each other, connected by and forming a bridge across the state line.

Once a wooden covered bridge, swept away in 1903, the Lambertville–New Hope bridge is now made of metal, painted a gentle green. A sturdy walkway on the downriver side of the bridge hosts foot traffic, but the road surface is a metal grid with three-by-three-inch squares halved into triangles. Vehicles crossing the bridge make a characteristic humming sound, and every Winter Festival and Annual Pride Festival (an LGBTA event), paraders uneasily eye the water moving directly below their feet. Within sight of that bridge, the Route 202 toll bridge, known locally as the "pay bridge" (or, by real old-timers, "the new bridge"), bypasses both towns. However, it is the metal "free bridge" that creates the bridge community, and Lewis Island lies directly upriver, with island visitors looking out at the bridge's pylons and up at the trusses and travelers that connect the two towns.

Perhaps as important in creating the bridge community are the hills that surround the towns. In Lambertville, Cottage Hill lies on the south side of town and is almost indistinguishable from Goat Hill in the more rural West Amwell Township surrounding Lambertville. To the east, Connaught Hill, also known as the Commons, was long considered a poorer part of town, and it has seen development in the early twenty-first century. On the northern side of town lies Music Mountain, named for St. John Terrell's Lambertville Music Circus, where big-name musicians played from 1950 to 1970.[1] Topographic maps reveal that these hills form the edge of the Appalachian range.

At the time of the American Revolution, when there were hills but no bridges here, both towns were known as Coryell's Ferry, named after the business that helped travelers continue on Old York Road (Route 202) connecting New York and Philadelphia. Most American schoolchildren know that George Washington's soldiers crossed the Delaware in Durham boats from Washington Crossing, Pennsylvania, to Titusville, New Jersey, as the two towns are now known, but in the Lambertville–New Hope area, some children can also tell you that Washington commandeered additional boats, including the Coryell ferryboat, for a second crossing farther north. Washington surveyed the river valley from a rock now known as Washington's Rock on Goat Hill, visible from Lewis Island.

In the nineteenth century, Coryell's Ferry split into two towns, although in both towns the ferry's location is still marked by a Ferry Street, and Lambertville salutes the old local family with a Coryell Street. When the Honorable John Lambert founded the post office in 1814, he called the place Lambert's Ville, but it was also known as Amwell at the time (Petrie [1949] 1970, 50–51). Locals—including the Coryells, of course—made known their feelings about the name change with the epithet "Lambert's Villainy" (Gallagher 1903, 27). Like in other places, however, bureaucracy in the form of the post office carried significant weight, and the town remained Lambertville, with its colonial roots and Victorian character. While Lambertville became known in hindsight as the home of James Marshall, the man who discovered gold at Sutter's Mill, California, and thus started the California Gold Rush, the town and its sister across the river, New Hope, contributed to the Industrial Revolution the rubber, lace, paper, and other mills powered by the Delaware. Of particular note to this study are the Union Paper Mill in New Hope, which employed most of the Lewis Fishery's early crew outside of fishing season, and the Lambertville Rubber Mill. The latter made Snag-Proof boots, which were given to the fishery crew, thus allowing the company to attest to high quality.

In the twentieth century, mill closings triggered a significant local depression, and in May 1965 Lady Bird Johnson, the country's First Lady, picked Lambertville as a place to visit during the Johnson administration's War on Poverty. Lady Bird's reflection on the visit in her book *A White House Diary* (1970) reflects a sympathetic yet superior attitude toward the less wealthy townspeople. She chose to visit the town in part because it was the second site for the Head Start program, and, in her book, she praises Head Start teacher Angelo Pittore (Johnson 1970, 310), whose family is still a well-respected resource for local educational and civic leadership. Johnson begins by putting a negative cast on the poorer neighborhood she visited, "called, for no reason I can understand, 'The Commons.'" The aside may simply reflect Johnson's unfamiliarity with the term, meaning a resource such as land held in common by the community (Hardin [1968] 2005), a customary feature of New England towns. However, her choice to call the name nonsensical leaves it to her readers to find sense, who in turn may interpret the word as meaning "low-class" or "crude." As she proceeds, Johnson seems to call on readers to think of the class-related meaning of "common" and perhaps embraces the insulting connotation herself. While praising a local church on the Commons because it "opened its arms to people in need," she also insults the neighborhood and its inhabitants, saying that the neighborhood has "a general appearance of shiftlessness"; she elaborates that she does "not want to turn America over to another generation as listless and dull as many of these parents looked" (1970, 310). One can only wonder about the extent of Lady Bird Johnson's influence on sensitivity about the name of the hill, which is now usually called "Connaught Hill" to avoid the insulting connotation.

Hard economic times did not end with Johnson's visit, and in 1972, the last of the Lambertville–New Hope mills, the Union Paper Mill in New Hope, closed. Fred Lewis, by then the well-known owner and captain of the Lewis Fishery as well as the mill's administrative superintendent, locked the doors. Locally, Lambertville was known as the poorer of the two sister towns; it was considered a bit rough and edgy, while New Hope garnered a more genteel reputation. Today, it is not clear which town is more "refined," as development in both towns has resulted in Lambertville acquiring more antique shops and art galleries, while New Hope hosts the chains such as McDonald's and Dunkin' Donuts. New Hope is still more widely known, but in 2013, *Forbes* magazine named Lambertville one of the fifteen prettiest towns in America (Giuffo 2013).

By the late 1970s, a gentrification process in Lambertville had started, with professional couples, artists, and other developers moving into the area, buying Victorian properties and restoring or renovating them. In the late 1980s, Fred Lewis served as a consultant to developers converting the Union Mill into condominiums. Lambertville's and New Hope's socioeconomic boundaries were then defined by class and length of residence, but with collaborations and friendships crossing categories. With the still well-known Lambertville and Golden Nugget antique flea markets on the south side of town, links to the Revolutionary War and California Gold Rush, and quaint New Hope with its historic "New Hope School" of Pennsylvania impressionism across the river, the hospitality and antiques industries grew in Lambertville through the 1980s and 1990s, drawing many weekend visitors from New York but also settlers, many of whom commute to Philadelphia, New York, or the New Jersey state offices in Trenton. Today one can see somewhat separate social circles of "old Lambertville" families and "new" families, with many shared networks and social spheres, such as the public elementary school and churches. These intertwining groups are separated by family history and, to some degree, socioeconomics, for the fortunes of many old-time local families are still affected by the local economic downturn of the mid-twentieth century, despite their hard work, while the newer families brought financial reserves with them to invest in the area. Disconnects and areas of tension occasionally surface, as is typical of areas where gentrification has taken place (Spain 1993). However, friendship, low-simmering resentment, and independent functioning are more often the norm than overt eruption along the old/new divide. Moreover, while not politically monolithic, Lambertville is a blue municipality in a red county, and a liberal, arts- and history-oriented small-town culture prevails, as is evident in the annual Shad Fest, itself a source of both communion and conflict.

Shad Fest

The Lewis Fishery plays an important symbolic role in Lambertville's award-winning annual two-day Shad Festival, first held in 1981. As folklorist Debra Lattanzi Shutika remarks, "at their core, festivals are multisense events that can express divergent, contradictory, or ambivalent meanings" (2011, 207), and Lambertville's Shad Fest is no exception. Almost always held the last full weekend in April when the shad season is usually well underway, the Shad

Fest can bring forty thousand or so people to the small town of four thousand for an event that includes a typical street fair but adds a heavy dose of fine arts and crafts, then grounds it in an environmental foundation story: the Shad Fest celebrates the return of the shad in larger numbers after ecological improvements made to the Delaware River took hold. The festival contributes significantly to spreading knowledge about the Lewis Fishery but also misleads island visitors into thinking that the fishery only operates during and for the festival. When I surveyed festivalgoers who came to see the haul in 2012, roughly 15 percent of respondents believed they "knew" that the crew only fished during the festival. Nevertheless, the Shad Fest promotes some of the fishery's themes (the "Big Stories" of chapter 2) as well as allowing the Lewis Fishery to tell its story to a wide audience of festivalgoers and people who see media coverage of the annual event.

General consensus credits a small group led by the late Jack Curtin of the Lambertville House (the historic hotel in the center of town) with conceiving of and driving the establishment of the Shad Fest. Muriel Meserve recalls that Jack Curtin

> really thought that after the fish came back that we should do something to mark this return of the shad into the Delaware, because that had been a long time in coming, and a lot of work on a lot of people's parts to get the river cleaned up. And it had always been, historically, such a big part of Lambertville's history. And he pushed and pushed until he got it through, and that first year they were so unsure that this was going to work that they didn't even put a year on the T-shirt they had made. (November 7, 2003)

From the start, the festival had a mixed focus, for Curtin drew a connection between the shad's return and the town's return to prosperity. Ellen Pineno, for many years the office manager for the Lambertville Chamber of Commerce, recalls: "Curtin said, 'If the shad can come back, then Lambertville can come back, too'" (Murphy 2007). Muriel Meserve remembers that, the first few years, "a lot of the people it attracted were mostly interested in fishing, but as it got bigger and bigger it became a real tourist attraction" (November 7, 2003). Today, the T-shirts are collected by visitors, townspeople typically own more than one, and people have taken to debating which year's T-shirt is the best. They are always dated now.

Like most folk festivals, the Shad Fest does not have singular roots nor a monolithic foundation story. Fred Lewis recalls that in 1949, when the town had a centennial celebration, Mayor Bill Naylor envisioned an annual festival for the town, which started in 1950 with the Liberty Festival; other one-time events followed thereafter. Then in the 1960s the town staged the Riverama Festival, which was held in the summer and focused completely on the river and river activities, such as a boat race. A parade and the crowning of Miss Riverama and her court were key features of the celebration, and similar events were also part of the Shad Fest very early on. Riverama involved many of the same people—or at least the same families—who later put on the Boat Club shad dinner at the Shad Fest from 1981 until 2004. The relationship between the two Lambertville festivals is not evident to everyone, and it's not surprising that people have different memories of the Shad Festival's start, since it took two years of planning before the first Shad Fest finally happened. One person told me that wife and husband Ellen McHale and Harry Haenigsen started the festival. However, in a 1996 interview, McHale recalled that Jack Curtin had recruited her and her husband to head up an arts and crafts fair as part of the festival, and they had drawn on arts events in New Hope in the 1940s for their inspiration. The confusion about the extent of McHale and Haenigsen's role is understandable, as some in town see the event starting as an arts and crafts festival (Murphy 2007).

The art-world focus of today's Shad Fest is the Shad Fest poster auction, which visually contributes to the fishery's efforts to tell what they're all about (these "Big Stories" will be discussed specifically and at length in chapter 2). The auction raises arts scholarships for college students from Lambertville and New Hope and is closely followed by locals and the arts community. Contributions from well-respected artists may draw prices in the thousands, while others made by community-minded hobbyists may kick twenty to fifty dollars each into the coffers. Poster subjects favor shad and other local subjects, and the fishery crew and scenes of Lewis Island usually appear in several posters. One painting of bagging up, the last part of the haul, by artist Mike Mann eventually became a tin-boxed jigsaw puzzle available at the town's game store. One year, the Best of Show, an acrylic portrait of Fred Lewis by Bob Beck, earned several hundred dollars in a bidding flurry that balanced regard for different aspects of the community. The fishery crew, in crew T-shirts, sat up front while Cap'n Steve bid, their connection to the painting's subject obvious to anyone with a real interest in the painting or the community. Other bidders drove up the price to honor the artist, then dropped out before the bidding reached the painting's actual artistic worth in deference to the family, whom those present apparently felt were the poster's rightful owners. To be sure, not every poster focuses on the Lewis Fishery, and several show little understanding of the fishery, or even of the shad, as guppy-like figures parade an any-fish-will-do attitude. Nevertheless, two of the eight categories for cash prizes emphasize the festival's namesake: the Shad Humor Award and the All about the Shad Award. Similarly, connection to the shad, river, community, and/or recycling have at times served as the criteria for the Shad Fest shop window decorating contest. There are also one or two booths selling shad prints made by putting paint on an actual fish bought at the Lewis Fishery.

The street fair may include some environmental organizations, such as River Keeper, but the nexus of the Lewis Fishery's ecological message lies on Lewis Island itself. The festival draws hundreds down to the fishery to see demonstration hauls using the traditional haul seine method, thus helping the fishery spread the word about its traditions and environmental efforts. These hauls are specifically called "demonstrations," because the Sunday catch must be thrown back as "demonstration only" (the fishery still follows an old blue law that prohibits it from fishing from Saturday 2:00 p.m. through Sunday midnight). The action educates the public about the fishery and at the same time further mythologizes the fishery. Because the hauls (even the Saturday haul, which can be sold if the boat leaves shore before 2:00 p.m.) are called "demonstrations" and linked to history in media coverage and official publicity, many viewers who do not stop to talk to the crew don't realize that the festival demonstrations are not historical reenactments but happen in the middle of the regular eight- to ten-week season. In any case, these public displays do raise the public's awareness of the fishery's activities, and many visitors do get the larger message, helped by scientific and educational agencies such as the Delaware River Basin Commission, which the Lewis family invites onto the island during the Shad Fest to educate the public about the river environment.

A Postgentrification Transformation

Joining "old" and "new" Lambertville, "newest Lambertville" can also be seen at the Shad Fest in the diverse crowd of locals. Historically, most local residents are of German, English, Irish, African American, or Italian descent. Starting in the mid-1990s after gentrification had taken root, Lambertville has become a "New Destination" (Lattanzi Shutika 2011) for Mexican immigration. Called by sociologists "New Destinations" because they are far from the Mexican border, these areas of suddenly increasing Mexican immigration resemble each other in that they offer similar work opportunities, particularly in the areas of seasonal work and the restaurant and hospitality industries. The gentrification of Lambertville at once built the hospitality industry and a population whose commuting schedules afforded less time for housekeeping and garden maintenance, but more funds to pay someone else to provide these services. At the same time, a miniboom of new construction in West Amwell and Solebury, the more rural municipalities surrounding Lambertville and New Hope, offered not only construction work but also the landscaping work to go with housing developments. The initial wave of Mexican migration then builds a second wave of opportunity for Mexican residents: serving as landlords and providing sources for traditional foods, including groceries and restaurants, some of which also attract the area's more cosmopolitan residents and weekend visitors from New York. For example, Tacos Cancun, a Mexican deli, was reviewed positively by the *New York Times* (Parish 2011). The Latino and Anglo communities are not yet fully integrated, but relations are not as contentious as those described by Lattanzi Shutika in *Beyond the Borderlands: Migration and Belonging in the United States and Mexico* (2011), in which she writes of Kennett Square, Pennsylvania, and Herndon, Virginia. The tensions of the early 2000s have become a rarity, while welcoming public sites and services (such as Adult English as a Second Language [AESL] classes and the Lambertville Free Public Library's Cuentos y Canciones, a Spanish-language story-time program) have aged into the norm.

Although this ethnographic section has seemingly wandered away from the traditional haul seine fishery topic, "the mix" is the point. Lambertville may at first look like the bucolic, isolated, small-town bubble that many nineteenth- and twentieth-century folklorists traditionally sought to study. And perhaps this is true, to some extent, if one ignores the connections the town's residents have to other places and people. More significantly, though, the seat of this ethnography is a varied social setting with intricate connections within the local space as well as global connections. The fishery crew and visitors have also been the people dining and volunteering at the Community Kitchen, the people coming to Lambertville as part of gentrification, the former mill workers, the public school teachers and students, those fixing and maintaining old houses, those running and patronizing the library, those surfing the internet and pulling in fishnets, and so on—all the time building relationships and caring for community resources through narratives. Thus, this examination of narrative may show how traditional processes help people in contemporary societies negotiate complex and changing cultural forces.

Who: The Key Players at the Lewis Fishery

Having traced the community history and demographics in the Lambertville area, let us now return and pick up the threads of the Lewis family and trace their connections to the community and the fishery (readers may wish to refer to the Lewis family tree in appendix B). For the family, the story always starts with William Lewis Sr., also known as Bill Lewis and Cap'n Bill locally, and as Poppy to family. Originally from New Hope, Bill Lewis Sr. began haul seine fishing in his teens by driving the horse that pulled the boat upriver. At this time, haul seining for shad was a lucrative enterprise, and in 1896 the local fisheries averaged about ten thousand shad each, each year (Fred Lewis, March 1, 1996). Bill Lewis Sr. fished in a total of five different Lambertville–New Hope fisheries, sometimes more than one at a time. He ran three of them, including one owned by Jonas Lear, his father-in-law. In 1888, at age thirteen, Bill Lewis Sr. started fishing at the Liberty Fishery on Holcombe (Lewis) Island. In 1918, he bought the southern half of the island and fished the two fisheries on the island, Point and Liberty, as one, which he called the Island Fishery and which is now called the Lewis Fishery. Bill Lewis Sr.'s sons, Bill Jr. and Fred, started helping out at the fishery early. By age five, Fred was spending most of his free time with his widowed father at the fishery, and in 1930, at age fifteen, Fred began filling in for men on the crew.

In addition to Captain Fred's father and brother, the key fishing members of the family include Theodore "Uncle Dory" Lewis (Fred's uncle), Edith "Eek" Lewis (sister), Nell Lewis (wife), Cliff Lewis (son), Muriel Lewis Meserve (daughter), Sue Lewis Garczynski (daughter), Donald Lewis (nephew), Steve Meserve (grandson), Pam Meserve Baker (granddaughter), in-laws, and others (a kinship chart can be found in appendix B). In Captain Steve Meserve's generation, his wife, Sue Meserve, and his sister, Pam Baker, have become cornerstones of the island, its fishery, and its community. When Pam returned to Lambertville in the mid-1990s with her husband, John Baker (also originally from Lambertville), and their two small children, Andrew and Sarah, they were integrated into the fishery, with John a regular member of the crew and Pam filling in for crew members and helping with selling fish and any other island tasks her mother and grandparents performed as owners. Today, Andrew and Sarah fish when work and college breaks allow—much as Steve did during his college years. Without knowing all these people, readers might feel that this opening curtain call risks becoming a sort of biblical begat list. Let me just underline here the fact that even in terms of who "the family" is, the lines of tradition-bearing are complex and intertwined, resembling more a postmodern mosaic than the classic, father-to-son line of occupational tradition so often celebrated.

One would expect that Bill Lewis Jr., about ten years older than his brother Fred, would have inherited and led the fishery after their father's death. However, Bill Lewis Sr. died in 1961 without a clear plan for succession. Both sons inherited the Lewis Fishery, and while both knew how to fish, neither had any training in leading a crew or the business. They did, however, have experience fishing together, as they ran a two-person fishery on their lunch hour at the Union Paper Mill in New Hope while their father's crew was fishing full-time over in Lambertville. Fred Lewis laughed when he described how he and his brother fought steadily through two whole seasons of leading the Lewis Fishery together. When his brother died, the fishery passed wholly to Fred. Fred owned the fishery until his death in 2004 at age eighty-eight, but around age eighty had to share the leadership because of his declining health. Again, the iconic father-to-son inheritance pattern was not to be, for both Fred's son, Cliff, and his son-in-law, David Meserve, predeceased him. Muriel Meserve remembers that after Cliff died in 1973, "it became very much my son [Steve], and he felt that from a very young age, I think, that this was what he was going to do; he would take over this fishery and preserve it" (November 7, 2003). Nevertheless, with Fred still able to participate in leading and not ready to give up the role, but not able to fish regularly either, the family needed a model of "taking over the fishery" other than simple inheritance. They created the position of "operator," and through it Fred passed the day-to-day leadership of the fishery to his grandson Steve, then in his thirties. At Fred's death, the fishery's ownership passed to Fred's wife, Nell Lewis, who became the fishery's first female owner, with Steve staying on as the captain and operator.

Today (2018), Muriel Meserve, now in her seventies, and her much younger sister, Sue Garczynski, own the Lewis Fishery and the lower half of Lewis Island, having inherited them from their mother in 2009. This situation continues the division of owner and operator roles established in the 1990s, sharing responsibility for the place and its traditions. Muriel and her siblings, Cliff and Sue, grew up on the island and in one or two other family houses in Lambertville. Sue, the first female ever to row with the crew, lives about four hours away in the Lehigh River region to the north, and she and her sons take part in the haul when they visit. Muriel, on the other hand, can clearly see the point of the island from her kitchen and living room windows.

It's important to say here that while their involvement with shad fishing eclipses all else in the media, the Lewis family is also known locally for their integrity, and their devotion to education and community. Local reporter Renée Kiriluk-Hill describes her working relationship with the *family* as one in which she trusts them as a *group* as an information source:

> Because I know the family, I think I came into [having Steve as my primary source after Fred's passing] figuring, well, Steve's going to be able to tell me anything I need to know, because I'm sure Fred passed it on to him. And he's also a smart guy like his grandfather [laughs] and like his mother [laughs]. So . . . I just

figured, well, when I ask him a question, he'll have the answer, or he'll get it for me, and that's another thing, is once in a while I'll ask something and he'll say, "I'm not sure; I'll get back to you," and, yeah, he does. (June 3, 2011)

To the reporter, the extended Lewis family's ability to pass on knowledge is paramount. Here, Kiriluk-Hill talks not only to the family's intelligence and knowledge but also to their reliability and integrity, which her career depends on.

The honesty Kiriluk-Hill expects is also related to the honesty that becomes a religious value for the family. Muriel, Fred, Nell Lewis (Muriel's mother and Fred's wife), and Pam Baker (Muriel's daughter) are all known in town for being "good Christians," a phrase connoting a combination of integrity, kindness, and a certain level of propriety. The family enjoys laughing at a piece of family lore that is ironic in retrospect: Nell's family was not too happy about her wanting to keep company with one of the Lewis boys and spend time over in Lambertville, but they were won over in the end by Fred's track record as an upstanding young man. The relationship between Fred Lewis, the island, and Christian spirituality will be discussed in more depth in chapter 3, but here suffice it to say that three ministers officiated at his funeral and clearly felt honored to be asked. On any given night, Pam or Muriel will be at the Centenary Methodist Church in Lambertville going to choir practice, attending a worship or administrative meeting, preparing food for the Community Kitchen, or preparing a children's event. Clearly, they are rocks in their religious community, but also add a sense of lightness and fun. Muriel Meserve is well known for her elaborate and creative Sunday School pageants and Vacation Bible School sessions, put on with help from her family.

The Sunday School activities can be seen in light of the family's commitment to education as well as to their faith. Muriel Meserve retired from several decades of teaching two—and in one case, three—generations of Lambertville Public School students (including her own children) and put in another several years as a substitute. In talk of particularly creative programs at the school, her name usually comes up as one of a talented, energetic, and well-loved team of teachers who developed scintillating programming before the word "interdisciplinary" entered the common vocabulary. As children, Pam and Steve were members of various school bands, but as adults they help with school performances and band camp. In such small school districts, extra hands are needed to help the students and teachers achieve their ambitious goals, and active participation from parents and other community members is a town tradition. Pam, Steve, and Sue (Meserve) volunteer with the student theater productions at the high school, drumming in the pit, building sets, and other tasks. Today, Pam works in the high school guidance department, inspiring students' affection through her special gift of identifying what is most important to them and showing interest so that students feel valued, not grilled. And of course, the family involvement with education began with Fred Lewis, who served on the elementary school board for many years, including those in which the town built the new, modern school. Taken together, this is a hefty family heritage of education, church, and integrity, a heritage that intertwines with the fishery work.

Traditional Haul Seine Fishing: A Primer

Having set the cultural, familial, and place-based backdrop, let us now go back and paint in the haul seine fishing tradition, which was once a common seasonal strategy for amassing a source of food or cash. The eighteenth-century *Pennsylvania Gazette* in Philadelphia contains several March announcements that inform crew members of various fisheries to meet at a particular place and time to start the fishing season. When I began fishing with the crew in the late 1990s, the call came not by newspaper, but by phone. Today it comes again through the written word: e-mail or phone text. According to tradition, the Lewis Fishery captain aims for the Saturday morning nearest Saint Patrick's Day, the first day of spring, or Nell Lewis's birthday. The specific date must be chosen each year also, depending on the river conditions, emerging signs of winter's end, and crew members' calendars. Historically, the shad season ends whenever the bulk of the shad seem to have passed Lambertville on their way upriver to spawn as far away as Hancock, New York, about 180 river miles north, which can be as early as the first week in May. The crew may continue to fish as late as June 10, according to the state-issued fishing license, which governs aspects unique to the traditional haul seine method, such as the size of the mesh that makes up the net, how far across the river the crew can go (they cannot stretch the net to the Pennsylvania bank), and which species of fish they can keep.

Community scholar Phyllis D'Autrechy's *Hunterdon County New Jersey Fisheries, 1819–1820* (1993) identifies dozens of fisheries from Trenton up through Hunterdon County, and Fred Lewis remembered six of them on the mile stretch of river around Lambertville in the early twentieth century, five of which his father fished at some point. New Jersey fisheries were known as the evening fisheries and Pennsylvania ones as morning fisheries, for the catch was best at the time of the day when the shad gravitated toward the bank the sun hit and thus favored one state in the morning and the other in the late afternoon. When Bill Lewis Sr.'s crew fished in the early twentieth century, captain and crew members took about a ten-week leave of absence from their mill jobs. The legal fishing week stretches from Sunday at midnight to Saturday afternoon, the week ending when the last boat leaves shore before 2:00 p.m. This law kept crew members' wives and ministers happy, as well as giving shad a window in which to escape the nets and spawn, ensuring the catch in future years. Bill Lewis Sr.'s crew might even fish round the clock when the weather and the fishing were especially good. Since the paths of boats and nets crossed the center of the river, crews would look across the river to gauge when their competitors were going out, and then take turns, delaying one's own haul if another haul had obviously already started. Crews would even race to be the first boat out on Sunday at midnight, since the first fishery with a crew in the water would likely get the most hauls that week.

Visitors often assume that traditions changed radically in the second half of the twentieth century and that the part-time status of fishery work came with the current, middle-aged crew's generation. Many are surprised to learn that part-time hours during fishing season became the norm as early as with Bill Lewis Sr.'s sons. Not regular members of their father's crew at the family fishery, Bill Jr. and Fred kept working full-time at the Union Paper Mill during fishing season, fishing at the Malta Fishery during their lunch hour, then walking back across the bridge to New Jersey to fish at the Lewis Fishery in the evening.

Fred Lewis's fishing career and the careers of those who came after him are colored by the supply and distribution situations. When Bill Sr. led the crew and fish exceeded orders, he sold shad to customers on the bank, then sold the excess in a few different ways. Farmers bought at the rate of forty dollars for a hundred shad, which they sold from their carts on the way home. Fish were also sent to New York markets via the train that stopped in Lambertville. Around 1935, when railroads were trimming their schedules, Bill Jr. and Fred ended their fishery day after the crew, trucking the excess fish down to the Philadelphia market late at night. With Fred's and Steve's crews, fishing after work and on Saturday morning for "extra" money, not a reliable income, became the norm. While the crew has always only kept what fish they knew they could use or sell, today's smaller catches always have individual buyers waiting. If the catch is good and the customers are plentiful, there may be two or even three hauls in an evening to fill orders, and lights may be set up on the bank after sunset. However, usually there is just one haul each day nowadays, crew members arriving most evenings around 5:30 or 6:00 p.m. and leaving the island around 8:30.

The fishing season and weekly schedule are still affected by river and weather conditions. Safety precautions keep a crew out of lightning or a high, fast current, and if the river is higher than the nets are deep, there's no point to making a haul and having the fish simply swim under the nets. The frequency of fishing is also determined in part by supply and demand, but even more so by today's contract with an intrastate cooperative to produce thirty evening hauls a season. Switching from a primarily catch-oriented operation to a data-oriented one has lengthened the season in some years, lasting weeks after the captain and crew believe that the bulk of the shad run has ended, but yielding valuable scientific information about the fish population. Sometimes a night's fishing will be held or canceled based in part on how many of the required evening hauls have already been made.

The Lewis Fishery has long kept records and samples for their own use and in cooperation with state or academic biologists conducting studies, including a ten-year tagging project in the 1970s and 1980s. In 2008, the Lewis Fishery entered a contract with the Delaware River Basin Fish and Wildlife Management Cooperative to collect data, benefiting Lewis Fishery, multiple state, and federal interests. Released from the previous requirement that they catch two hundred shad each season to keep their grandfathered license (the state stopped issuing licenses in the 1940s), the Lewis Fishery values the stability the contract gives them, which enables them to preserve the tradition of haul seine fishing. From the cooperative's point of view, the government gains access to data gathered for over a century by a rather consistent method, with samples and particularly robust data since the 1990s. For each haul, the fishery records the time the boat leaves shore, temperature of the water, height of the river, length of net, and weather conditions, and then counts and records how many fish of each species are caught. For each shad kept (some are thrown back if they are considered "too small" and thus "worth more in the river" as part of the reproduction cycle), the crew notes the sex, length, and weight, and then takes a scale sample. Someone records each shad's data on a small manila envelope containing the scale sample. A shad's scale shows the shad's age, like rings on a tree. During the journey northward, the fish does not eat but rather lives off the minerals in its own body, thereby changing the appearance of the scale: each raised ridge shows a spawning year. Studying the scale can also tell the scientists what the fish has been eating and information about the river itself.

The method of data collection has stayed virtually the same with the new contract. In the inaugural year of the contract, one of our first contract inspectors looked dubious as one of the fishery kids, then twelve, stepped up to take measurements and record data on a haul. The inspector praised her work, and his surprise increased when we told him she already had five years' experience with the job. It may have been a new contract, but the collection work had gone on for all of that crew member's life. Like Bill Lewis Sr.'s counts a century earlier, this data still gets logged into the daily ledger, along with sales. Despite the passage of 125 years, mounting industrialization, and fluctuations in the shad population, traditional haul seine fishing at the Lewis Fishery has stayed pretty much the same over five generations.

Early haul seine fishing histories in this area frequently mention that the Native Americans taught the European settlers to fish (see, for example the Shad Fest '99 insert in the *Lambertville Beacon*), although that was by a somewhat different method using a "battery," which Fred Lewis describes:

> Well, originally, the colonists didn't have the nets over here. The way they did it originally, they—the Indians showed them how to catch them—what the Indians would do, especially whenever shad were spawning, they would put a row of stones around and leave one end open, so the fish would run up in there and they'd close that end off. And then they'd go in and catch the fish out with their hands. They do it in shallow water.... Another way they would do it is they got into the sections of the river with a mud bottom. They would pile stick brush down in there and make fences around and then close it up. And what they would do is, they would have this pen. And they would go down the river from the pen, and they'd take the brush and beat the water to chase the fish up in there. And that's how they got the term they usually use to identify a commercial fisherman ... a bushwhacker. (March 1, 1996)

Fred reveals that he has been called a "bushwhacker" himself and notes that in the fisheries below Trenton the fishermen also referred to their nets as "bushwhackers." One can see the Lenape method described and illustrated in anthropologist Mark Raymond Harrington's novel *The Indians of New Jersey: Dickon among the Lenapes* (1938), a work still sold at the New Jersey State Museum in Trenton. However, nineteenth- to twenty-first-century methods and tools also resemble methods used traditionally by Europeans. Rather than building batteries and moving the fish into the enclosure, the nineteenth-, twentieth-, and twenty-first-century Lewis family crews maneuver the nets around the fish out in the river, then pull them in to land. A haul can last an hour in a low river, or it can take fifteen minutes with a short section of net in a high, fast river.

Basically, the haul begins when the crew loads the net onto the boat. The net comes in 50-, 100-, 150-, or 200-yard lengths, about ten feet deep at most, which are combined in relation to the river conditions. The net must be adjusted in response to each six-inch rise or drop in the water height. The captain determines how much net to use, how to maneuver it, and whether the river is too high to fish at all, based on a gauge on the footbridge, or by noticing which familiar stones on shore can be seen above the water's surface. During his retirement, Fred Lewis could add or remove net sections or mend net before the crew arrived to fish. Today the crew loads the net into the boat while Captain Steve and another crew member mend, lengthen, or shorten the net—but still using the same hand-carved net-mending needles, now seventy-five years old. Stopping to repair a "crow's foot," "box," or "zig-zag," three names of typical tears, creates pauses during the load, which are filled by conversation and enjoyed—except when the crew's patience is tried by a particularly holey net, bad weather, or obligations scheduled after the haul.

Years ago, Bill Lewis Sr. made his netting through a process called "knitting," although any needleworker can tell you that the process is not at all what is usually called knitting, but is rather *knotting*. Fred Lewis paints a picturesque scene of his father's off-season activity:

> He used to sit there all winter long. He'd be knitting net. And he'd have to knit, aw, he'd knit about a hundred yards—would take two hundred yards of netting to make a net a hundred yards long.... He would knit about a yard of net in an hour.... We used to live in New Hope on Mechanic Street—and we had a window that looked out on Mechanic Street. And he would sit there in the evening, and after dinner, he'd knit until midnight, and just sit there and watch the street. Or some of his old cronies would come in, and they'd sit there and talk while he'd knit net.... But then it got so that he couldn't see very well, so he bought the nets ready-made then. (March 1, 1996)

Fred's description resembles the Northern Irish *ceilis* described by folklorists Henry Glassie (1982) and Ray Cashman (2008), occasions when men would socialize informally in the evenings.

Today there are similar scenes, but with a mixed-gender audience on the riverbank during the net-mending stage. Now most of the nylon netting itself has been purchased, although there may be several sections where considerably large patches are handmade during the mending process. These lengths of net are then "hung" using a series of half-hitches to fasten the netting to the lead line and cork line, which hold the bottom of the net down and the top of the net up, respectively. The construction of the "corks" and leads themselves reflect the eclectic, do-it-yourself nature of the fishery. Not actually made of cork, the corks are about six-inch-long synthetic beads strung on the cork line, with the occasional homespun cork fashioned from a two-by-four with its edges rounded off. The leads' creation shows how old-school the Lewis Fishery gets. Steve Meserve fires up the wood stove in the crew's cabin, melts lead in a large ladle, then pours it into a century-old mold to form two-and-a-half-inch-long bars. When these cool, Steve gets out his hammer and anvil to flatten the ends, bend the length around the rope, and then pound the flattened ends against one another to secure the lead, thus creating the lead line. Lastly, the net is fastened at the sides to two brails, which are long poles weighted at the bottom that help hold the net open and give it a firm end the crew can manipulate with a rope.

Like the net, the boat is retrofitted, because, again, almost nobody does what the Lewis family does at the fishery. As Steve Meserve says, "There's no standard-issue oar for the boat we use, or there's no standard-issue net the way we fish. It's all pretty much handed down, and you adjust it with the materials you've got. A lot of what we do here is whatever you can get your hands on" (April 22, 1996). His wife, Sue Meserve, a carpenter like Steve's great-grandfather, Bill Lewis Sr., chimes in, "Well, you guys have a tradition of you're not going to hire somebody to come in and do it for you. [Steve: Right.] You don't have the money to hire somebody to come in. You're going to do it hit or miss. Well, we can make it work, probably pretty well, and, you know, a little more technology gets added now and again" (April 22, 1996). Steve also notes, "We do a lot of things temporarily that last a long time" (April 22, 1996). This seems to be a corollary to Steve's other maxim, "My family is English and we keep everything," a statement that explains why almost-century-old, half-finished or warped hand-carved net-mending needles sit in the drawer next to the needles still in use.

When several fisheries dotted the banks in the Lambertville–New Hope area, the crews used wooden boats built locally specifically for the purpose: the hull was large enough to hold the net in the stern and still leave room for two or three rowers and one person pushing the boat in with a long, metal-tipped pole. After a boat was stolen in the mid-1960s, the fishery had no funds to replace it with a traditional boat and started the first of the retrofits: a boxy, fiberglass-bottomed, US Army boat with handles on the side for portaging and an Odyssean saga of its own. In 1956, the US Army Corps of Engineers began planning the very unpopular Tocks Island Dam project, which involved evacuating and flooding many homes—pointlessly, since the project was eventually abandoned in 1975 (Albert 2002). With periodic flooding, one of the Army Corps' boats escaped multiple times and floated downriver, to be retrieved by the engineers. Eventually, the engineers forsook the vessel, and it sat in a farmer's field filled with rocks until the Lewis family heard of it and bought it for use as the fishery's new boat. To convert it to a fishing boat, they simply removed the engine to make room for the net, added a false bottom, then retrofitted oarlocks. The oars themselves are made from two-by-fours that are easily replaced when the river breaks them. After decades of patching and repatching the fiberglass bottom, the fishery finally gave up on this boat. The second nontraditional boat turned back the clock on a brand-new fishing skiff. The crew never installed the motor to begin with and again retrofitted the hull with a wooden false bottom and oarlocks. The new boat offers a little more room than the Army Corps' boat did but was unofficially christened the SS *Lardbutt* in recognition of its heavier weight, which can be felt by the crew throughout the haul.

After the crew loads the net onto the boat, one person gets into the boat to steer, accompanied by children who are more easily supervised this way. Years ago, a horse pulled the boat up the riverbank, but today the crew uses a rope to pull the boat upriver about a quarter of a mile, give or take, depending on the height of the river, net length, and weather and path conditions. One person takes the rope attached to the land end of the net and walks it down the riverbank, fulfilling the role of landsman, with the aid of a special leather-and-metal harness called the "smalick." While two people row the boat out into the river, the third person in the boat minds the net, making sure that it "pays" (slides) off the back into the river, not catching on anything and tangling. Eventually, they turn "toward home," meaning back to shore. With a combination of rowing, drifting, and poling, and with the captain and landsman communicating about when the landsman should "hold back," or walk downriver, the right shape to the curve of the net emerges. Once the sea end lands back on shore downstream from the land end (which is still being walked down the riverbank), the cork line makes a *C* shape, with the riverbank connecting the two ends. The sea (southern) end is then walked downriver to the spot where the net will be pulled in and coiled, while the land end gets into final position. "Bagging up," the last part of the haul when both ends of the net are pulled in simultaneously, usually happens at the point of the island but may also occur on the side of the island if the water is low, or if the crew is demonstrating for Shad Fest visitors. During bagging up, most of the six-plus-member crew will be on the creek side of the island point holding the sea end of the net, with several people on the cork line and one or two people on the lead line. The hardest task is "first leads": pulling in the sea-end lead line with someone else coiling the net behind him or her. The shorter land end is pulled in with the captain on leads, the landsman on first corks, and another crew member coiling the cork line behind the landsman.

It's not until the very last moments of the haul that crew and visitors can see if anything is in the net. Everyone leans forward, hoping especially so catch a glimpse of American shad, which today are about eighteen inches long, give or take. Considered a "pretty" fish, this member of the herring family has a profile with sleek curves, a slender body, a silvery sheen, and a characteristic line of about four charcoal-colored spots, beginning larger at the shoulder and stretching below the dorsal fin, seeming to fade out. During bagging up, the shad catch the evening sunshine and flash. Muriel Meserve describes the final moments this way:

> The excitement when they bring in the nets; I don't think that will ever change. When you hear that flapping, and people—no matter how old they are or what year it is—are going to be right there looking, as soon as you start to see that quiver in the water. And that hasn't changed a bit. It doesn't really matter if it's one fish or ten, or what the fish are, even, just the idea of them appearing out of that water, I guess, doesn't change. There's an excitement of seeing that net come right into shore, with something, anything, in it. (November 7, 2003)

A cry of excitement arises spontaneously as the water starts churning and splashing, and the fishes' reaction creates more excitement. When stressed, shad exude a slime to help them elude predators, and when the net comes in, slime, scales, and water fly everywhere.

Then another flurry of work starts. The captain and the woman leading sales confer about what numbers and types of fish are needed to fill customers' orders. The fish are then given a preliminary sorting. Those to be kept are put in buckets to take under the cabin, and the others are thrown back. The fishery's license distinguishes "food fish," which can be kept and sold, from "sport fish," which must be thrown back immediately. Line fisherman among the visitors groan audibly when beautiful striped bass (some a yard long), trout, muskies, and large- and small-mouthed bass happily slip back into the river. Food fish typically kept and sold include shad ($6 for a roe/female and $4 for a buck/male), catfish ($1), carp ($5, $10, or $15, depending on size), quillback ($2), and suckers ($1). Considered almost worthless, the gizzard shad—a bonier, smellier, uglier, worse-tasting cousin to the American shad—are usually thrown back, along with a few other species. While the fish are sorted, some crew members hold up the cork line to trap the fish splashing in shallow water, while other people throw back fish, musically calling out the name of each type to the person—often a girl—taking the count on a tally sheet. While not requiring as much strength as the more prestigious rowing and first leads tasks, taking the count (being the girl with the clipboard and pencil) is one of the most difficult jobs, for it requires good hearing, concentration, and speed. While throwing the fish back, crew members notice whether any fish need help reorienting themselves to swim out into the river, and occasionally landsman Ted Kroemmelbein will be seen gently pulling a stunned fish back and forth in the river to help fill its gills with extra oxygen until it heads off with a purposeful splash.

After the count, most of the male crew members interact with the visitors as they pull the net back from the river or load it onto sawhorses, depending on the expected schedule for the next day. There are always visitors on the bank who are not customers, most of them locals appreciating their town tradition, people from farther away who come to see the fishing as a "rite of spring" and family tradition, folks interested in science or history, line fishermen, inspectors, and occasionally diners or tourists who heard about the tradition through Shad Fest media coverage or just happened on the activity while on a stroll. The fishery crew always pauses long enough to show children the live fish, because it so delights the little ones and because crew members want children to connect with nature and with their food sources. As Sue Meserve says, food doesn't just come from the grocery store (October 3, 2009). Meanwhile, the women and girls, sometimes with the help of another member of the crew, carry the buckets up the bank and under the crew's cabin, which serves as the fishery's market. First, any customers wanting only fish other than shad are taken care of and sent on their way. Next, the shad data are collected and recorded. Then the fish are simply wrapped in brown paper and put into grocery bags.

While selling the fish under the cabin or chitchatting on the riverbank over the years, I have learned many food traditions that affect which fish species and preparation methods are used by particular segments of the customer base. The fishery's visitors and clientele make up an ethnically diverse group, and white European Americans, the predominant group in the local communities, were the largest group in the past. They are the most likely to make reference to the shad-planking traditions found in other regions. In this area, however, they bake or grill shad and value the shad roe above all, sautéing the roe sacks or frying them with eggs. African Americans are the second-oldest group to be included among the fishery's clientele but are more likely to plan meals with fried shad and also frequently ask for catfish. The well-established central New Jersey Asian Indian, Bangladeshi, and Pakistani populations share many shad-cooking customs. A similar fish exists in India, called the *hilsa* or *bing*, which is curried or steamed with banana leaves. Shad are also presented to hosts as an Indian hospitality ritual. Some Chinese customers want shad heads and tails to make soups, and they make suckers into a paste that they preserve and use to flavor other dishes throughout the year. Most recently, members of the Hispanic community are discovering that they are welcome to visit Lewis Island and have access to fresh fish there. However, so far they only occasionally experiment with buying fish at the fishery, being accustomed to different species of fish that are available in local grocery stores.

While their contract with the intrastate cooperative now drives the Lewis Fishery's economics, fish sales remain important to the fishery's primary purpose, because sales provide income and the opportunity to educate the public about the tradition and environmental issues (see chapter 2). However, the overall drop in profits following the decrease in the shad population has essentially changed fishery economics on the employment side of the equation. Traditionally, gross profits would be divided by ten, four shares for the captain (for owning the net, the boat, and the fishing rights, and for being the captain). If the captain also fishes as a crew member (rather than just directing the action), he can clear half the profits to support his family, who, in the old days, fed the crew as well as sold the fish. Minimally, a fishery the size of the Lewis Fishery needs six crew members, but a haul can be done under certain conditions with as few as five, and now a larger crew ensures having enough crew members on any given night. In the past, the magic number of six signified maximizing profits by splitting into fewer shares, but as time went by and crews grew to eight or nine members, the Lewis Fishery captain took fewer shares for himself. Steve Meserve recalls that when his grandfather fished with the crew,

> it was still a good way to make money, not the kind of money my great-grandfather was making relative to the salaries of the day, but it was still, you know ... Johnny [Isler, a landsman who retired in the late 1990s] would talk about being able to take trips to California on the money he made fishing and stuff.... I bought my first drum set with my fishing money. So, you know, other kids would go work at stores and gas stations, you know; this is what *I* did for working. (October 3, 2009)

In the late 1990s, some crew members were considered to be "full" or "regular" crew members who received a full share—maybe a hundred dollars or two—at the end-of-season picnic. Other crew members who were not part of the rowing rotation or came less frequently might receive a fraction of a share. Adults who came only occasionally and kids who didn't row were not part of the share system. Over time, these shares dwindled with the shad population, and crewing has been a volunteer proposition since around 2003.

Instead, the profits are intrinsic to the values and purposes of the Lewis Fishery as the last of its kind. Crew members, customers, and visitors may acquire shad, build muscles, or accrue volunteer hours for Scouts or school. More importantly, though, is their participation in the tradition, the season, the environmental project, and the social community. In this landscape they tell, build, and become parts of stories, and thus take care of their communal resources. This system of narrative stewardship begins with the large, overarching stories that inform purpose, which we now turn to as the subject of chapter 2.

TWO

Fishing with Purpose: The Big Stories

I got started, because my father was into it. And he began fishing back in 1888 . . . and ever since then, and one of us has been doing the fishery ever since then. We never missed a year.
—**Fred Lewis** (March 1, 1996)

In these three sentences, Fred Lewis tells perhaps the most important story, the biggest Big Story, about the Lewis Fishery and Lewis Island: it has persisted for a long time through a family's faithful work. Another time-anchored Big Story is the story of incongruous springtime, with its unpredictable weather that dislodges the dead branches and grasses of the previous year, yet also brings new life, activity, and warmth by the end of the season. There is the Big Story of preserving the fish, the island, and the river. And lastly, the family's mandate to share the island forms a cornerstone of the Big Story of promoting civility and community. Together, these stories create a common purpose for Lewis family members, the fishing crew, customers, and other visitors, which creates a certain intensity on the island.

First, a Word on Genre

When I say the "Big Stories," I mean something in the sense of the old phrase, "What's his/her story?" The answer here would tell what the person is "all about," or what the person's backstory or motivations are. This is similar in scope to the body of narrative known as "myth" in folklorists' classic division of oral narratives into myth, legend, and folktale (Ben-Amos 1992, 101–2). In scholarly discourse, myths and legends, but not folktales, are told as true with varying degrees of literal, factual truth. While *legend* adds more details and "purports to be true" (Ben-Amos 1992, 102), myths relate more to representative truth.

Folklorist Linda Dégh's three-part system divides narratives into "tale genres, legend genres, and true experience stories" (1995, 60), including various types of myths within the category of legend (1995, 76–77). A particular

Big Story can easily ride the myth/legend divide or the legend/true experience story divide (Dorson 1973, 110–11), and so I prefer to use the term "Big Story," which allows including stories with the detail typical of legend or even personal experience narrative but which also reaches the scale of myth's abstract meanings and symbolic import (Geertz 1973b, 127). While Big Stories may have distinct characters and plots, they also hold much in common with what literary analysts call a "theme," a unified message that a work of literature communicates, either overtly or by implication (Booth and Mays 2011, 246–49). Returning to conventional folk narrative typologies, a myth can be said to embody or communicate a theme. Similarly, even a personal experience narrative can be shared as a way to tell a related Big Story.

Big Stories at the Lewis Fishery also have much in common with the way Clifford Geertz interprets Balinese cockfighting as "a story [the Balinese] tell themselves about themselves" (1973a, 448). Both Balinese cockfighting and traditional haul seine shad fishing are akin to and interrelated with both ritual (enactment) and myth (narrative). At the Lewis Fishery, people fish as a way of telling their Big Stories, and tell their Big Stories—to themselves and others they invite into the tradition and place—about fishing and while fishing. Moreover, the idea of the Big Stories goes beyond narrative in the conventional sense to include activity, following Geertz's view that "cultural forms can be treated as texts, as imaginative works built out of social materials" (1973a, 449). Considering the collection of these forms, Geertz says, "the culture of a people is an ensemble of texts, themselves ensembles" (1973b, 452). Thus, the Big Stories are not simply narratives but collections of "cultural forms" or "texts," which in turn interact to create a complex and subtle value system on Lewis Island.

Telling the Big Stories

Folklorist William Wilson divides storytellers into three categories: "situational," "conscious cultural," and "professional" (cited in Stone 1998, 26). Situational storytellers relate narratives in the course of casual conversation, while conscious cultural storytellers purposefully preserve group culture by preserving the stories and passing them on. While the fishery family and crew (especially the captains and schoolteachers among them) are most likely to act as conscious cultural storytellers when they tell the Big Stories, almost anyone showing up at the fishery may relate some part of a Big Story as a situational storyteller—the situation being a visit on the riverbank. The professional storytellers most likely to tell Big Stories are reporters, photographers, and documentary filmmakers.

The main ways of learning the Big Stories come to a head during the annual Shad Festival, which brings hundreds of visitors to Lewis Island to watch a haul, often drawn by media coverage starting in the week leading up to the Shad Fest. When told as discrete stories rather than as parts of other stories, the Big Stories appear in a number of contexts. Big Stories show up several times each week on the island during fishing season, and during the rest of the year in various locations. This story type also figures in classrooms at the local school as part of science or social studies projects. Muriel Meserve, who taught upper-grade social studies at Lambertville Public School, notes that at the school "the shad always ended up in *something*," meaning some lesson or other activity, and "the perfect time to do it was always around Shad Fest" (October 5, 2009). After all, she stated on another occasion, "the town is based on the river, and we can use the fish as an example for a lot of things" (November 7, 2003). The Lewis Fishery has also told the Big Stories in adult education situations, not only my own cultural anthropology course at Bucks County Community College (Bucks County, Pennsylvania, lies directly across the river from Lewis Island), but also in the anthropology classes Fred Lewis and Steve Meserve visited at Raritan Valley Community College (Lambertville's community college), lectures they gave to environmental organizations, and the like. In particular, Nell Lewis tells of her husband's special fondness for meeting with fishing clubs:

> Yes, Fred used to go to some fishing clubs. He'd get invited to a fishing club, and speak, and of course he [would ask]—"have any questions?" And when he gets to talking, and they'd keep him there for a long while. They always loved that, asked questions, and would hear stories from him, and . . . he used to love to tell about fish. It was *about shad*. (October 5, 2009; emphasis in the original)

The Lewis Fishery's Big Stories about shad have even been told at the 2004 Smithsonian Folklife Festival's *Water Ways* exhibit, live, in the festival program (Belanus 2004), and in online exhibit materials (Smithsonian Institution 2004).

Less overtly, the Big Stories are enacted. In other words, Big Stories are "told" through the *actions* of the fishery crew, Lewis family, and visitors to the island. While the fishery family and most of the regular crew are aware to some degree that they enact these Big Stories when they do yet another haul or adjust their home and work lives to commit time to the fishing season, they are most conscious of the Big Stories when explaining the fishery to outsiders. Visitors also participate in narrative activity through their vital role as audience members. Later, they retell the fishery's Big Stories, incorporating themselves into the story when they transform their visit to the fishery into a narrative frame. Audience members' awareness of the Big Stories is perhaps clearest when they are overtly told a story as part of the "why": why this fishery is important to the earth, why this tradition is important to the regional culture, why this family and plot of land is important to the community. Here, narrative stewardship at the fishery has much in common with ritual, which I define as an action or set of actions that function symbolically. The Big Stories both highlight and embody the coherence that connects actions (e.g., fishing, observing, telling) and values (e.g., environmental preservation, tradition, sharing). At times, the embodiment is literal as the crew enacts the story by fishing, which is different from simply "acting out" the fishing, as a historical reenactor might do. The story of tradition is re-created, not just rehearsed, through the activity of fishing, just as the story of community is re-created as well as retold through people gathering on the island and having friendly, face-to-face conversation.

Outside of introductions to the fishery traditions and telling through enactment, though, the Big Stories may also be told through other stories, perhaps as a reflective coda on the end of a story. Muriel Meserve notes that the fishery has "always been kind of a teaching operation . . . either teaching or research, one or the other [laughs]" (October 5, 2009). Similarly, this metadiscursive comment by Pam Meserve Baker within a story points out greater significance to the listener: "[The fishery's] part of us, and as long as we can do it, we need to do it, because it is something that people need to be cognizant of" (October 4, 2009). With this comment, Pam points to the fishery's Big Stories of education, specifically about the environment. Conversations between the fishery crew and visitors on the bank may reinforce newspaper articles or documentaries about the Lewis Fishery, because conversations connect to the bigger themes, the Big Stories, that are broadcast by the media.

Media coverage of Lewis Island and the Lewis Fishery may be local (anywhere from Hunterdon County to Trenton to Bucks County) or regional (central New Jersey or Philadelphia), and at times has been as "big" as the *New York Times*, PBS, NPR, and the Travel Channel. The festival coverage has appeared annually in local and regional newspapers and television news programs as far away as Philadelphia, and at times warranted a full supplement in the *Lambertville Beacon*, Lambertville's longest-running newspaper (1844–2015). After having been assigned to cover the fishery in the advance festival coverage, *Trenton Times* photographer Mary Iuvone now seeks out the fishery at the festival, which, she feels, *makes* the festival: "I think [the demonstration hauls are] the coolest part of the whole festival, really. You know, 'cause without that, it's just another festival. You know, it's another street fair" (May 20, 2011). However, the perennial festival coverage and the perennial fishery coverage do not overlap exactly.

Reporter Renée Kiriluk-Hill, who covered the fishery for more than twenty-five years, most notably for the *Hunterdon Democrat*, the county newspaper, reveals an almost ritual pattern: "From having done this a number of years, I have a pretty good idea now [when] ... the shad should start running, so in the back of my head, I'll say, 'It's time to call Steve Meserve and find out, is the crew going out yet?'" (June 3, 2011). At the start of spring, Kiriluk-Hill checks with the captain to see whether the hauling has started and prints a short article announcing the start of the season—itself a sort of rite of spring that she has executed for about twenty years. Kiriluk-Hill works a bit more tightly with the fishery family to build her story than the average reporter does: they are on a first-name basis, and she has even had them provide a picture for her article when a photographer is not available. Nevertheless, her process resembles the activity of other reporters in local papers, most notably Joe Hazen, whose family owned the *Lambertville Beacon* for four generations (1869–1989). In addition to editing the paper (Mack 1989) and writing regular newspaper articles, Hazen also wrote the popular Chit-Chat column, which ran for many years. Marking the start of the season, Hazen's April 2, 1998, column reads: "When Fred begins his springtime work, we must recognize that it is time to change the spark plugs in our lawnmowers. Go ahead, put the snow blower away, even though it didn't earn a rest." In this column, with its casual tone, Hazen names Fred Lewis by first name, showing familiarity with the man and assuming that the audience will know what Fred's "springtime work" is. In years when shad provided a regular part of Lambertvillians' diet or activity, an update or two might appear during the season, sometimes alongside advertisements for shad prices in local grocery stores. This notice from the May 4, 1888, *Beacon* commands notice today, for it hails from the year Bill Lewis Sr. first ran the fishing at the Point Fishery, the site of today's Lewis Fishery: "Reports from our local shad fisheries lead to the conclusion that those engaged in the business will not be likely to become millionaires from this season's profits."

Although people frequently think of the media as being outside the folk community, older *Beacon* articles provide an emic view of the Big Stories, because the Hazens and the Lewises *are* local history (the Hazen family acquired the *Beacon* in 1869, about twenty years before Bill Lewis Sr. began leading his family fishery). However, when the regional Packet Publications, known first for its flagship paper, the *Princeton Packet*, bought the *Beacon* from the Hazen family in 1989, the connection between the paper and the town stretched and thinned. Joe Hazen worked for Packet Publications for about another decade, but he was sent to other Packet towns to work, and the *Beacon* office moved to neighboring Mercer County and frequently supplemented its pages with Princeton news before ending altogether in 2015. The fishery coverage, which was a story in itself for the Hazen *Beacon*, became more of a small piece of the Shad Festival story for the Packet *Beacon*. The media's Shad Fest coverage creates a kind of shorthand that connects the fishery tradition with the festival, thereby helping fold the festival into the phenomenon of narrative stewardship. The coverage, like the fishing itself, is a perennial part of local storytelling practice. Present-day media coverage is dominated by its connection to the Shad Fest, which has existed for only about a quarter of the fishery's history, but the media nevertheless plays an important role in telling the Big Stories to the general public and becomes a part of the fishery family's strategy for telling the Big Stories.

Intertwined Big Stories

The environmental Big Story intrinsically connects the fishery and the festival, because the Shad Fest celebrates the river's and the fish's rejuvenation. However, media coverage also tells the fishery's other Big Stories as well. Aware of the perennial nature of her work with the fishery, Kiriluk-Hill says that she tries to revolve different stories:

> I'd either look back in our files or I'd recall, well, I focused on the family history last year, so this year I'll focus more on the environment, because it gives the reader something new. Because the readers *do* look year after year, but then every few years, you have to refresh something that you may have written previously, because they've forgotten. (June 3, 2011)

In terms of the Big Stories, this means that she alternates emphasizing the environmental Big Story, the Big Story of tradition, the Big Story of springtime, and sometimes even the Big Story of civil community. The Big Story of civility is not *covered* as often as it is *experienced* by the press. That is, while the press benefits from the Lewis family's cultural values of respect, service, and sharing, the photographers and reporters who most successfully cover the story are those who best follow the ideals of Lewis Island. Success here is judged by full and accurate coverage, fishery family response, and/or the amount of work the fishery people do *for* the media. Perhaps the most successful media person working with the fishery at this point is Renée Kiriluk-Hill, who is one of the few of the media whose name Steve easily remembers and whom Pam Meserve Baker feels comfortable talking to. While greatly respecting the work of Fred Lewis, Kiriluk-Hill was one of the first to recognize the leadership of Steve Meserve and built a partnership with him so that she could achieve their mutual goal: informing her readership of what is most interesting, accurate, and valuable about the Lewis fishery. Echoing Kiriluk-Hill's words, Muriel Meserve says of teaching the shad's story in the public schools: "That's the thing . . . you could do something different *every year*" (October 5, 2009). For example, one year the students built a working model of the river across the front of the stage in the auditorium, going from Hancock, New York, the shad's most northerly spawning ground, down to the Delaware Bay. Another year they used bulletin boards in the hallway to show the Delaware River islands.

Told in a variety of performances—both narrative and ritual, representative and literal—the Big Stories act as the skeleton of narrative stewardship, giving structure and strength to both stewardship and the storytelling system. Kiriluk-Hill sums up the richness of the narrative stewardship system when she addresses some of the Big Stories:

> [Shad fishing in Lambertville] is an age-old rite, and, along with that, there's the shad bush that blooms, a native American plant, that blooms when the shad is running. It's part of the connection. And, of course, what is the Lewis Fishery about? *Connections.* Community connections, connections with the environment. It's . . . and what is Lambertville about? You know, Lambertville is very much what was, not just what is. And Lambertville *honors* what was, so the Lewis Fishery is a *big story* in this area, because of that. (June 3, 2011)

Here, Kiriluk-Hill addresses the connections between the stories she writes about, which approximate the Big Stories discussed in this chapter, including the story of the connections between people, which will be discussed below in the section on the Big Story of civil community.

When Steve Meserve describes "the story" of the fishery, he also weds the three stories of ecology, spring, and tradition, which has a family connection for him:

> Well, for me it's a personal, family history. It seems like it was always part of the family, because my great-grandfather started it, at such a young age running it, and then buying the property and continuing it, being stubborn enough to keep fishing even when there were no fish, and having my grandfather continue it, and then myself coming along to also continue it. And the reasons changed over the years, and that's, you know, my great-grandfather started it, because he could

make good money in the time that the fish were there, but then [the purpose] changed to being more of a social indicator of a passage during the year, like leaves falling in the fall, ... but the shad returning to the river was a sign that spring was here and all its celebration that that brings. And we were the ones to help signal that. Without a presence like ours, the fish can come and go pretty much unnoticed. (October 3, 2009)

Many years of providing the story to reporters, visitors, curious coworkers, and others have enabled Steve to create an incredibly condensed multitopic Big Story, but in this chapter, we need to unpack the interrelated assemblage. Despite the challenge of pulling the stories apart, an introduction to each of the Big Stories will compose much of the rest of this chapter, identifying each major theme and exposing the underlying values.

The Environmental Story

The Lewis Fishery's family and crew have a role in protecting and preserving the shad, the river, and the island, and in terms of narrative stewardship, they do this through the environmental Big Story. This Big Story marries science and tradition, two fields frequently pitted against each other in popular thought but that fit together easily in relation to education and conservation, two key interests of the extended Lewis family. When relating how shad and the fishery entered the local elementary classroom, Muriel Meserve says they focused on

> usually the environment. And that goes back to my grandfather, too [as did the fishery's founding], because he lobbied everybody under the sun for cleaning up the river, back before, you know, environmentalism was even a word, but he kept saying, "if you clean it up, they'll come back," and of course they did. (October 5, 2009)

This story relates to Bill Sr.'s haul ledgers, a business tool that recorded numbers caught and purchased. Begun in 1890, the books also served as a data source allowing him to track fluctuations in total shad counts that went beyond the normal cyclical changes for shad and indicated to him an alarming drop in the shad population. A good way to trace the change is to compare averages over a few decades. From the 1890s through the 1930s, the average totals per year for each decade range from about 1,600 to 3,000 shad. In the

1940s, the average total per year drops to triple digits with just over 700. In the 1950s, Bill Sr. confirmed his concerns when he continued to lead his crew despite the dwindling counts. The average total shad per year for that decade is under 20 shad, including two years with 0 shad. Knowing that the number of hauls fluctuates each year, one might compare the average numbers of shad in each haul, which the fishery started tracking in the mid-1920s. Here, the general trend of plummeting numbers in the 1940s and 1950s is duplicated. The average shad per haul for the 1930s is 5.4 (rounded up to the nearest tenth); for the 1940s, 3.8; and for the 1950s, a depressing 0.8 (see appendix C for a full list of shad catches from 1890 to 2017).

Many in the family tell the story of when Bill Lewis Sr. wrote letters to everyone he could think of and testified before a federal commission meeting to get government help for the shad and the river. In verbal shorthand, Pam recounts hearing her teenage daughter telling the environmental Big Story at the Shad Fest to Jon Corzine, then governor of New Jersey, and then begins to act out the Big Story herself:

> She got most of it right, or darn close.... The 1888, doo–doo-doo–doo-doo. We yada yada, yeah, the pollution block, da–da–da–da-da cleaned it back up. You know? The letter writing that went on, you know, and the certain terms that we use, you know, ... what does Mur say? I don't think it was, um, "tenacity"; I think it was "orneriness."[1] You know, just the, you know, keep going, 'cause it, you know [voice becomes dramatized as if speaking multiple times to different audiences], you have to clean this up; we have to clean it up; well, I think you should clean this up; well, you need to clean this up. (October 4, 2009)

Here, she's acting out the repetitiveness of the letter messages to various authorities, changing her voice slightly with each repeated directive to artfully highlight the persistence: the message is the same, but there are several utterances and audiences over time. In this "Mr. Lewis Goes to Trenton" story, progress comes but comes quite slowly, as his testimony plays a role in establishing an intrastate commission, which discovers a pollution block in the Philadelphia-Camden area. Once the pollution block is cleaned up significantly, the shad begin returning up the Delaware in increasing numbers starting in the 1960s, according to the fishery's records of catches and family stories about the crew's and townspeople's excitement at seeing larger catches.

Kiriluk-Hill observes that the newspaper stories tell the environmental story more completely over time, building as our society learns more about the environment and as the readership expects more:

> Obviously, we've come back to this time and time again in stories: the pollution block at Philadelphia that really put an end to the shad runs, at one point. And as, again, time goes on, people have become more understanding of *why* things like this happen. So, we're more comfortable putting out—not just [the] pollution block, but *oxygen*, you know, getting a little bit more detailed *because* . . . our knowledge of [our effect on the environment] and our coverage of it has changed over the years at large. You know, not just at the *Democrat*, but when you see any newspaper, you see more specifics. Not just "Oh, it's global warming!" Not just "Oh, it's water pollution!" but more details. (June 3, 2011)

In Kiriluk-Hill's recounting of the environmental Big Story, she refers to the pollution block that the intrastate group found after examining the river at Bill Lewis Sr.'s urging. This story is "the first chapter" of the environmental Big Story. Bill Lewis Sr.'s persistent cry of danger becomes a repeated cautionary tale in the fishery's environmental Big Story.

With the return of the shad in larger numbers, the Lewis Fishery's day-to-day work has not changed significantly, but the focus has changed from earning a living to stewarding natural and communal resources. The data recorded from each haul has become detailed, but then, continuous careful observation is one of the fishery's traditions. Over the years, Bill Lewis Sr.'s love for scientific study has been repeated as various biology doctoral students, the state Department of Environmental Protection, and intrastate organizations have relied on the fishery to collect reliable data on the catches, including the contract in effect today. For about ten years from the mid-1970s to mid-1980s, the Lewis Fishery took part in a tagging operation, run first by a federal agency, then by the corresponding state body now called the New Jersey Division of Fish and Wildlife. According to Fred Lewis, in this project, the fishery tagged and released half the catch and sold the other half, and Ted Kroemmelbein recalls that crew members were paid for a later environmental project. In addition to allowing scientists to track the movement of individual fish, these projects enable the government to calibrate a sonar counter north of town that estimates the shad population on the Delaware (Fred Lewis, March 1, 1996). In a classic "win-win" solution, the fishery's traditional methods produce the data that scientists need, and collecting data enables the fishery to preserve the regional fishing tradition. The change to a contract was in the way the work was framed, not in the way it was done.

While the environmental Big Story has become more and more important to the Lewis Fishery's purpose, it has also been told more prominently through the growth of the Shad Festival. One key and consistent aspect of the Shad Festival, usually appearing in yearly media blips and features, is the foundation myth that the festival celebrates the return of the shad up the Delaware River in greater numbers after the river was cleaned up. The Lewis Fishery had an instrumental role in creating the circumstances (the cleanup and increase in the shad population) that the festival celebrates, but during the festival the story of Bill Lewis Sr.'s active role is usually only told by the fishery crew on the island. Others at the festival point mostly to the fishery's activity as a local shad-related tradition. At times, the Lewis Fishery seems almost taken for granted, in the best sense of the phrase. Nevertheless, the historic haul seine fishery supports the festival's bid for authenticity by intrinsically connecting the festival to place ("You need to come to *our* town to experience this")[2] and anchoring it with the weight of history. The Shad Fest's environmental Big Story easily blends with both the Big Story of tradition and the Big Story of spring.

Big Stories of Time: Tradition and Spring

For many in the Lewis family as well as those interested in local history, the greatest of the Big Stories is the story of tradition: the longevity of shad fishing in this particular spot by this particular family; how the tradition lasted so long and why that's important. In recent years, either a speaker or an audience member may make the connection between the Lewis Fishery and John McPhee's award-winning *The Founding Fish* (2002), some listeners even aware that the Lewis Fishery appears in that book. However, the fishery's Big Story of longevity was told long before McPhee discovered it and continues to be told, most often in local conversation or in newspaper articles without reference to the popular book. The story of the Lewis Fishery as a tradition is often triggered and represented by the stock phrase of the fishery's claim to uniqueness: it is the only traditional haul seine fishery on the nontidal Delaware and has been so since 1943. This phrase is usually coupled with another fact, here in Fred Lewis's words: William Lewis Sr. "began fishing back in 1888 . . . and ever since then, and one of us has been doing the fishery ever since then. We never missed a year" (March 1, 1996). Narrators underline the continuity by mentioning 1953 and 1956, when no shad were caught in the nets all season. The other most important date in Lewis Fishery history is 1918, the year that William Lewis Sr. bought the southern half of Holcombe Island. After leading the two fisheries on the island separately for six years, he began to fish a stretch of bank overlapping both so that they merged into one named Island Fishery, which we know today as the Lewis Fishery. This basic story kernel then may move backward to connect haul seine shad fishing with

Native American practices or horizontally to connect the Lewis Fishery to the other now extinct haul seine fisheries along the Delaware or to shad netting in other rivers.

Photographer Mary Iuvone's appreciative response to the Lewis Fishery's longevity is fairly typical: "It's, you know, this sort of old tradition that hasn't died, and that . . . even in this twenty-first century, it still happens. I mean, I think that's pretty cool. You know, there are not too many things out there that are still done maybe the same way that it was done like a hundred years ago" (May 20, 2011). When Kiriluk-Hill tells the Big Story of tradition, she refers to this angle in her Lewis Fishery newspaper stories as the "history," which, like the environmental story, she deepens over time with more information and details:

> History is another part of it that we'll go back to again and again and that we consider important because the Indians taught the seining methods to the colonists. They showed them how to build the pens. Again, we go back to that to show the continuity, the link, and how it's been adapted or hasn't. You know they're still going out there—you [corrects herself to include me as a crew member] still go out there in a rowboat! [laughs] . . . So . . . a lot of it is explaining cause and effect, history, the link through time, even more recent history one hundred years ago when there were numerous fisheries. You know, every once in a while, we will bring that up, that the Lewis Fishery was just, oh, another one, [if] you go back to the early part of the twentieth century. Today they're unique, but again that's how things have changed. (June 3, 2011)

Here, Kiriluk-Hill hits the major points of the historical story: the Big Story of the tradition passing from group to group, and also from generation to generation of Lambertvillians.

The Big Story of tradition often ends with reflecting on why this fishery has had staying power. By 1943, the Lewis Fishery was the last commercial haul seine fishery on the nontidal Delaware. Also in the 1940s, the shad catch started to dwindle such that the IRS reclassified the Lewis Fishery as a hobby in 1947, and it then became more of a "sideline" for the family (Fred Lewis, March 1, 1996). The Lewis Fishery certainly had a landscape advantage. While only line fishermen in boats can get to the spots by the bridge supports that shad favor most (Pfeiffer 1975, 97), fishing on an island enabled the Lewis crews to fish in higher water than the other fisheries could, because they could pull in net on the point of the island at water levels where other fisheries would have already had their net pile up on the riverbank. The historical narratives, however, emphasize human behavior. Some fisheries dropped out because of drinking and fighting among the crew, two activities not tolerated at the Lewis Fishery: granddaughter Muriel Meserve says to her mother and me that Bill Lewis Sr. "did run a tighter ship, I think, with his crew than a lot of them. [Nell Lewis: Oh, yeah.] I mean, they did have beer over here [at the house on Coryell Street] . . . and they would do *some* drinking, but not like some of the others" (October 5, 2009). Fred Lewis was even more demanding. Again, Muriel recalls, "Dad wouldn't even let [the crew] put their beer in his refrigerator and, you know, there were things that that's just the way it was" (October 5, 2009).[3] It's important to note that Fred's rules were followed not because the crew and family feared him but because they greatly respected him (Steve and Sue Meserve, April 22, 1996). The Lewis Fishery's longevity can be attributed to several Lewis family traits: having high standards for crew behavior; maintaining high standards for handling the fish (pulling them in over cement or felt so they wouldn't taste muddy, not grabbing the fish by the gills so that the heads don't turn red); stubbornness (determination to do it every year and not let the license run out); and the ability to stretch financial resources with a tradition of "making do" or jerry-rigging whatever is needed.

Like the boat, the footbridge to the island has also been adapted and recreated for generations, resulting in another Lewis family saying: "We've built more bridges than Roebling." Here, the speaker references the famous nineteenth-century builder, John Roebling, associated with his bridges and namesake town, Roebling, New Jersey, just over twenty-five miles downriver. Much to the puzzlement of any casual visitor unacquainted with the river's inconvenient flooding habits, the bridge is designed to rise with the river and creek levels, breaking into preordained pieces that are tied to each other and to trees on both sides of Island Creek. These pieces then float in place until someone on the bridge building crew mounts each of the two pieces like a raft, disassembling decking and joists while crawling backward on the still intact piece. The work crew then reerects the trestles, reassembles the deck and railings, and lastly ties everything together, ready to rise and break with the next flood. One bridge lasted as long as eight years (1996–2004), although others have lasted only from a given spring until hurricane season just a few months later. With so many objects handmade, objects and their use become attached to local people, not mass production. The people then become tradition bearers of the place and the activity, and of the narratives told about both.

The Big Story of tradition is particularly welded to the progression of captains from William Lewis Sr. to William Lewis Jr. *and* Fred Lewis, to *only* Fred Lewis, to Steve Meserve (a trajectory that will be discussed at length in chapter 3). Suffice it to say here that through each challenging time of passing the traditional captain role, the fishery's cultural norms of civility and hard work were key as the crew responded to change. For example, Fred's "old crew" morphed by ones and twos into a "new crew" (of which I am a member), most of whom have always known Steve as a leader and never as the "kid" on the crew. In fact, only a couple of us were fishing back when Steve led the crew *with* Fred. Those most knowledgeable about the fishery's history, however, have a still older crew as a point of reference: Bill Sr.'s crew of his brother, Theodore ("Uncle Dory") Lewis, and coworkers at the Union Paper Mill. These older members figure in now-classic *Newark Sun* photographs from 1937, and, looking at their roster, one sees some familiar surnames such as Bachorne and Stout, as their direct and indirect descendants make up the active community in Lambertville, and in a couple of cases more recent crew. Similarly, there are generations of "fishery kids," children of the fishery family and crew, many of whom became regular crew members, starting with Bill Jr. and Fred. Particularly at the start of each season, conversations among family, crew, and visitors share information about happenings since the previous season, observing the changes and continuity in the generations.

For most, one of the Big Stories of the fishery is its part in the story of the turning wheel of seasons: the story of spring. Visitors to the island and media coverage perennially connect the season, the place, and the act of haul seine fishing, and I like to joke that the reporters and photographers appear each spring alongside the wildflowers and ducklings in the week before the annual Shad Fest. The connection between fishery and spring has become cultural, but it is literally natural. Independent of humans' fishing activities, shad feel compelled to return to the waters where they hatched, and simultaneously the shad bush (or serviceberry) blooms, responding to the same temperature triggers. Once we bring nature into the cultural world, however, the shad's spawning becomes a "sign of spring" or a "harbinger of spring," as so many people say, and doing or watching the haul becomes a "rite of spring." Kiriluk-Hill connects this rite of spring to shad fishing history: "It was a *very* important food source through the centuries, and, so, the running of the shad was a pretty exciting [laughs] time. Fresh fish! We're not eating whatever's dried and left over" (June 3, 2011). Mary Iuvone calls the fishing a "rite of passage for spring" (May 20, 2011), and her conflation of the anthropological phrase "rite of passage" with "rite of spring" marries the single season within the annual cycle to any person's life, in particular the point of adolescence and fecundity.

Iuvone's words are readily echoed by visitors on the island, who profess coming to watch a haul as a ritual of experiencing spring. As Pam Baker says, "They make it part of their yearly ritual too, just like the shad" (October 4, 2009). Muriel Meserve says emphatically, "It really is a rite of spring.

You really feel that spring isn't here until you've gotten the first fish and you know things are moving along. I don't know; I can't imagine spring anywhere else. And I missed it when I was away [at college]" (November 7, 2003). Here, Muriel references the most overtly ritual-like custom at the fishery: recognition of the first fish. Although usually more than one shad appears in the net in the first haul that includes shad, one is named "the first fish" and is given to someone specially. In Bill Lewis Sr.'s day, it would be given to the general manager of the Union Paper Mill, where he and the crew worked. Later it would be given to a special customer, such as Mr. and Mrs. Hentschel, a sweet couple who came from a distance every year and seemed to really understand the sense of grace and civility expected at the fishery. Later, the fishery gave the first fish to a coworker of Sue Meserve's who was the great-grandson of the general manager of the Union Paper Mill, because the symmetry of repeating the interfamily custom three generations later was so delightful. As will be discussed in depth in chapter 6, noticing signs of spring is intrinsic to the haul itself. The Big Story of spring draws together people in activities that are more or less spiritual, and more or less secular, which leads us to the Big Stories about how people should act toward one another.

The Big Story of Civility and Community

As one comes across the footbridge onto the island, one sees almost immediately a handmade sign that reads, "DO NOT THROW THE STONES."[4] The sign is so iconic that a man once tried to buy it as a wedding gift for his bride to commemorate his successful marriage proposal there. The sign's message is intended to be taken literally, but also has metaphorical import, and perhaps the bridegroom wanted to set a positive tone for his marriage. Indeed, even with its authoritative tone, the "DO NOT THROW STONES" sign plays a part in conveying the Big Story of civility and community, including the family's mandate to share the island. While anyone is welcome on Lewis Island, it's an "anyone comes" but not "anything goes" kind of place. On Lewis Island, people are expected to behave civilly, share, and, well, not throw the stones.

The literal cause for the sign enlists visitors in caring for the space they have entered when they step off the footbridge. The Lewis family protects the island from the river, which perpetually threatens the land mass with erosion and perennially makes good on the offer. Sticks? Sticks are fine to throw. It's not that the island family wants to squash fun. Heck, family and crew members skip stones with the best of them—but not on the island. There, every stone is needed, and some are dragged from other places to help "hold the island down," particularly in the flood and dusty seasons. Moreover, if thrown back into the river at the point of the island, the stones might tear the nets.

When I interviewed crew and family members about the island "rules," each mentioned the "do not throw stones" rule, and each had experience promoting and complying with the maxim. Muriel Meserve remembers her father's struggle promoting the rule for decades:

> The [rule on Lewis Island] that everybody in *town* knows, I think, probably, is you don't throw stones! [laughs] And what goes with that is that somebody has to haul all those stones here, and they've been, you know, smoothed out to keep— that it's smooth through the nets, and you can't be picking [the stones] up and throwing them out there, no matter how great it looks [laughs]. And I think Dad has explained that to every kid in town at some point . . . which is what he would do with it, I mean. Unless they were repeat offenders, he would always explain very carefully why you couldn't do that: "I know it looks like fun, *but*" . . . On the other hand, if it's the third or fourth time, he'd just get really mad [laughs]. (October 5, 2009)

And when he got "really mad," Muriel Meserve continues, "usually he'd just go down and tell them very strongly, that, you know, 'you have to stop.' Ah, once or twice I saw him tell somebody, 'just leave,' you know, if they would start talking back to him, but very rarely" (October 5, 2009). One should note that, like son Steve and daughter Pam on other occasions, Muriel goes from metadiscourse (talk about the story) to ritual activity: she starts telling a story *about* saying something, but then switches to reenactment. While this narrative activity attests to the Meserves' love for theater, it also shows the intrinsic connections between *telling* and *doing* for them as narrative stewards.

Attempting to throw a stone from Lewis Island also acts as a shibboleth, separating insiders of various gradations from newbies: (1) those who know not to throw stones without reading the sign, (2) those who read it and then comply with the expectations of the place, and (3) those who can be told or can read the sign, and still choose to follow selfish whims, this last group becoming subject of surveillance stories (see chapter 4). The double message of the sign was perhaps most obvious in the years when the first reflective *S* sticker fell off. Even with the *S* rewritten by hand, the sign seemed to say, "DO NOT THROW TONES," dictating (ironically) civility in times when the freedom to be rude has been safeguarded. From the late twentieth century and into the twenty-first, "civility" has become a concern on college campuses and in workplaces,[5] but promoting high standards of communal behavior has been one of the Lewis Fishery's Big Stories for over a century.

Civility and environmentalism entwine in the directive to leave the stones on the island and a similar Lewis Island rule: to not disturb the island's plant life, either. Muriel Meserve says, "That was kind of an unwritten thing: ... You can pick a few [flowers, but] don't dig anything up, you know, don't take it off. It's not going to live anywhere else, so you have to just leave it here and this is where it belongs. And [it's] pretty much that way about everything over there: just leave it ... the way it is" (October 5, 2009). Here, Muriel indicates the importance of civility on the island: not only do you not throw stones or pick flowers, which endangers the island and disrespects the work of the people who hauled the stones there, but you also do not argue with Fred Lewis when you are a guest on his property. Fred Lewis, for his part, tries not to lose his patience, raise his voice, or send people off the island for not behaving appropriately.

Bill Lewis Sr., and even more so Fred Lewis, insisted that the Lewis Fishery crew behave well, even a little better than the other fisheries' crews, when those existed. Steve Meserve inherited a crew and business culture in which members refrain from excessive drinking and fighting while working. More than ever, being civil, even kind, was the norm. From a teenager's point of view, crew member Keziah Groth-Tuft describes island etiquette this way: "You don't act like a brat." Her description of the dictate clearly reflects a fourteen-year-old's expression, but the view matches those of adults whom I interviewed: "Well, you can't complain. Everyone has a job; you gotta do it. You can't complain about it. You can't be fussy like that. The whole 'don't throws stones, 'cause it ruins the island,' and you have to respect people. There's more respect for your elders, whatever, it feels like there [on the island]" (Keziah Groth-Tuft, October 3, 2009). Recalling a year when Fred Lewis suspended a crew member for the rest of a season because of his behavior toward other crew members, Muriel Meserve concurs that there's "a sort of unspoken [rule] how you treat each other in the crew" (October 5, 2009). She has trouble articulating the social rules precisely but clearly feels that the standard is long standing and characteristic of the Lewis Fishery: "I'm sure that's a carry-over from all the way back. And I don't know that anybody ever actually put it into words. And that's one of those unwritten things. There's a line. You're okay to *here*, but there's a line that you can't cross" (October 5, 2009). She goes on to admit that there was "a lot of, you know, swearing and yelling ... there ... [laughs], with some of the crews that they've had, but still there's that line. You know, where you can't get personal with somebody else, you can't, you know, attack personally, that ... just don't do that" (October 5, 2009). The last sentence, "just don't do that," echoes the firm tone in Muriel's descriptions of her father teaching visitors not to throw stones or molest the plant life. Yet, while here she indicates the line, literally outlines the minimal standard of behavior, the Big Story of civility often aims higher at the marks of sharing and kindness.

Sue Meserve emphatically recalls that she was immediately struck by the remarkably good nature of the fishery crew, being accustomed to working with predominantly male crews in theater shops she supervised or represented as union president:

> I wasn't used to a group of men who were as nice as these guys were. There's no girly pictures in the cabin. It's not allowed. While the more we carry on, the more profane we get, Fred's a Christian, and you don't, you don't badmouth too much. You don't use too many words[6] here. We may be up there [on the riverbank where the boat goes out] and start really swearing, but we're not going to swear down here in the cabin. (April 22, 1996)

Avoiding swearing and girly pictures shows the men's inclination to curb traditional male behavior for a higher ideal of civility. Ted Kroemmelbein explains that crew members limit swearing on the island, particularly in front of children, because they "represent" the fishery (June 15, 2013). Sue also often looks back to a story involving children—traditionally a woman's purview—to show the male crew's quality: "Somebody—Johnny's granddaughter—was missing one night. They didn't know where she was after Girl Scouts, you know. Well, all these guys up—I mean, they were ready to mount a, a patrol and go out and find this child! And I just never experienced guys like that" (April 22, 1996). This is not to say that the crew members are Pollyannas in hip boots. There is certainly a fair share of teasing and occasionally mild swearing. However, Keziah also described the people on the island as "nice and kind" (October 3, 2009). Clearly, she and Sue both consider the fishery crew to treat others in a way that is better than average, and they appreciate being treated better as well.

Although management methods used with the crew have changed somewhat from when the crew members were employees, not volunteers, the Lewis Fishery has always been anchored in a sense of fairness and respect for each member. For example, even before I became a full member of the crew, I was given a half share to recognize my work with the women under the cabin, which may not have been on par with the labor of a full crew member rowing the boat but was nevertheless a considerable commitment. Moreover, for each haul and from day to day the captain fairly rotates harder positions (first leads, rowing) and easier positions (cork line, second leads, catching the net), but also with consideration for a particular crew member's injuries, preferences, or scheduling issues. This spreads down to crew members' deference to each other—such as when tenth-grader Zach stepped aside from pulling leads to allow my daughters to do this favorite task when an alleviation of their school obligations allowed them to fish at the end of the 2017 season.

The Big Story of community and civility is also enacted daily through the system of sales. My first clue that the sales method was special at the Lewis Fishery came during my first year of observation, long before I was a crew member or fishwife. I was shyly holding my baby outside and peering under the cabin while the women distributed fish to the customers. A middle-aged woman, who had earlier gone under the cabin to buy with her elderly aunt, emerged looking sheepish, and in her embarrassment playfully confessed to the infant something along the lines of, "I got in trouble and got sent out here, so I'm going to stay here with you where I can't get in the way anymore!" Later, I followed up with the fishery women and learned that the older woman had indeed ejected her niece so that she could conduct business. The younger woman was interrupting the conversation to urge her aunt to pursue getting the larger fish or more fish, which was inappropriate to the Lewis Fishery sales system so well known to her aunt. Knowing that she could not quickly teach her niece such a complicated system that balanced social values and relationships with finances, she declared her an outsider—literally—to spare the Lewis Fishery and herself more bother and embarrassment. It took me many more weeks to understand the basics myself, and years to understand how it has adjusted in response to changes in the customer base and fish supply.

At the Lewis Fishery, fish are priced by the type of fish and, if a shad, the sex of the fish (the higher price going to the egg-laden roe shad). In the late nineteenth and early twentieth centuries, shad might also be categorized by whether it was a "giffen," a shad under three pounds, and priced lower (Fred Lewis, March 1, 1996). This word has fallen entirely out of use, because the shad population has shrunk in both number and size such that "giffen" would describe much of the haul, thereby making the distinction irrelevant—if the fish is *really* small, it is thrown back, not sold. Unlike at most fish markets, there is no price per pound, and perhaps even more unusual is that sales are not on a first-come, first-served basis in the usual sense.

While there is little competition for suckers, quillback, catfish, and carp, there are usually not enough shad (not enough roe shad in particular) to give all customers the full number desired. Therefore, the sales method developed around the prevailing concept of sharing as fairly as possible, with both sellers and customers playing respective roles. By showing up on the riverbank (or, for many years, phoning the head seller during the day), customers get themselves on the daily list and register what type and number of fish, including which sex of shad, they want. The sellers do not just keep filling orders until they run out, as is expected in many businesses. When the haul comes in, the head fishwife, interacting with the other fishwives and the customers, decides who gets which fish, trying to make sure everyone gets something, at least, and as much as possible of what they want, if not everything. The first person may not get everything he or she wants, nor does the first person on the list get all the biggest fish. In making her decision about who gets what, the head fishwife will balance size (big is preferred), sex (female is preferred), and number. Of course, no deal comes out perfectly, and a customer's place on the list, commitment, and generosity come into play to determine who gets more of what they want or the best-looking fish.

Sue Meserve describes how the women use the rotating list and other values in concert to make sure that the shad are distributed as fairly as possible.

> Your name goes on the list. If all the fish come in on one haul and there's not enough for you and you say, "well, I'll, OK, I'll just take two." Well, then, on the second haul [that night], you don't stay on the top of the list to get the rest of your fish; your name goes to the bottom of the list and we give somebody else a chance because you've got some. You know, and that's the best thing that we can go by. But, it's like, no, we have a list; it's first come, first serve. Put your name on the list, you know. If somebody else wants to defer to you or say, okay they can have mine, or something, it depends on how Grammy [Nell Lewis] feels. She may say "well, ok, that's fine," or she may say, "well, no, no, I don't, you can't really do that, because there's lots of people on this list here. I mean, if you don't want this, alright, but I got another name taken before you can give it away" to a guy, you know, like number fifteen on the list or something. (April 22, 1996)

In addition to the names that get put on the list each day, the fishwives often have a mental list of when they expect long-term customers to arrive, who may receive deference because of their age or the distance they must travel to faithfully keep the tradition. Notwithstanding the *Hägar the Horrible* cartoon posted in the cabin in which the Viking crew complains of being paid in "dead fish," there is also an oral list of crew members who may need a fish for a particular purpose, such as providing a shad dinner for a visiting relative. Crew and family defer to customers as much as possible, and the fishwives will take that also into consideration when balancing meeting the needs of the various customers with those of family and crew. Again, they keep in mind needs, the order of the list, showing consideration for others, and commitment.

Commitment comes in the form of coming to the fishery and being physically present at the time the haul comes in, capital that can build over a number of weeks or years. Someone who is present under the cabin at fish selling time will move up the list ahead of people not physically present, which is important when fish are few. Customer 1, who is present, higher on the list, and planning to return the next day, might even choose to defer to Customer 2, also present but lower on the list. Customer 1 will likely receive some consideration for the second haul or on the following night. When necessary and feasible, the fishwives may also show commitments of their own when they cannot make up the difference with fish. For example, while they usually offer the assistance of someone on the crew to wheel a messy or large order over the bridge and load it into the customer's car, only on rare occasions for a particularly kind or long-standing customer might they clean a fish. The long-standing customers who understand the values of civility and commitment embedded in the selling system may use that knowledge to inform their own behavior. For example, Sue works regularly with one customer who has come for years, buying fish regularly to freeze, because he does not eat meat. If there will likely be multiple hauls in an evening, he will defer until later hauls, when the crew is likely to "fish it differently," adding more net, going farther into the river, and perhaps getting more fish, or perhaps catching nothing at all. Sue feels that he follows the system "right": "And he said to me, you know, like, I'd hold up a fish to him, because I know him and I'll do it, I'll do it out of courtesy to him and he'll say, 'you know best, you know best, you know what's there.' You know, I've handled all the fish" (April 22, 1996). She recognizes and appreciates the way he shows respect for her opinion and trusts her. In considering his behavior as a customer, she draws a connection to how nice the men on the crew are, conjecturing that this customer might be one to jump in and help if one of the kids fell in the river (April 22, 1996). Here, the customer/crew line blurs to show how people build relationships and live the Big Story of civility on Lewis Island.

Perhaps more importantly, though, the value of generosity is told and lived through the Big Story of the family's mandate to "share the island" itself. When Bill Lewis Sr. bought the island in 1918, he wanted to share the resource with the community and insisted that his family do the same. This sharing is not absolute—after all, it is private property with a private home on it for which the family has financial and legal responsibility—and so the family does reserve certain places and times on the island for themselves and their friends. Yet, these "private" gatherings usually occur only in one part of the island (around the house), while outsiders still can wander around the rest of the island. The degree of sharing routinely practiced on the island by the extended Lewis family is very unusual compared to most other private property in town—or even in the United States. Muriel Meserve recalls her father saying that his father always felt that, when he let everybody come over that it was to be shared. You know, you didn't close things off and keep them to yourself—it was almost an Indian kind of, you know, attitude toward it. And then if you believe that, then you also don't take things away from it, you leave it there [Charlie Groth, confirming understanding: for the next person], m-hm. You make it just as enjoyable for them. (October 5, 2009)

Here, it is clear that the standards of behavior on the island (e.g., not throwing stones or moving plants around) are directly linked to stewardship roles: the Lewis family protects the island so that they can share it with visitors, and visitors must do their part in the stewardship process so that the island is around for sharing well into the future. As Muriel Meserve's daughter Pam Baker says, her family "just wanted [the island] to be used. We're all together; we need to share this space" (October 4, 2009). In this statement, Pam moves from the concept of private property to the shared human condition.

Muriel Meserve reflects the value of the island as a shared resource as she recounts the many nonfamily activities on the island over the years. While she calls it "a *family* party place" (October 5, 2009), and the family *does* reserve parts of the island for private use, the island is *not solely* for *her* family:

Yeah, there's *always* been that sense that it's a part of the community, it's there for the community to use, and, it's a place that people choose to ... mark events in their lives. So it gets a lot of use for weddings and memorial services and baptisms and, you know, the high points, and it's always—I don't think anybody's ever been turned down [chuckles] for one of those things. You know, there've been an awful lot of people that have asked to use it for that or for a religious service that's important to them. The synagogue has used it for different services, and there have been, you know, sunrise services and vesper services [laughs], and just about everything else over there. (October 5, 2009)

To the list of events held there, Pam Baker adds youth-oriented gatherings such as Halloween parties and sleepovers of her friends, even as young adults, and young missionaries "crashing" there over the summer (October 4, 2009).[7] Daniel Garber, of the twentieth-century New Hope School of Pennsylvania impressionism, tells the Big Story of sharing the island in his painting, *Lambertville Beach*, which depicts the bank of Lewis Island filled with sunbathers and waders on a typical summer day (Garber 2018; James A. Michener Art Museum 2018). To be sure, in the past many people learned to swim in Lambertville Recreation Commission lessons on the banks of Lewis Island.

Muriel Meserve says that one of her "favorite stories about sharing with the community ... is the miniature golf course":

> Oh! This is just the neatest thing! And this is just so typical. Dad and Uncle Bill [Bill Lewis Jr.] during the Depression built a miniature golf course. This was before the house was built, right? It was back up there, in back of where the house is, up toward the woods, and they apparently built this whole, I mean, this is before I was born so I, this is strictly from stories. But this was great that they did the whole thing. They strung Christmas lights around, so you could play at night, and the whole thing, and their *idea* was that they were going to open it up and they were going to let people play there for like, I don't know, a penny, nickel, whatever. Well, it was during the Depression. Nobody ever had any money. As Dad said [laughs], nobody ever paid a penny to [laughs] play there, but the people from town played on it all through the Depression. [laughs] (October 5, 2009)

As Muriel told the story and noted the expense of replacing golf balls, her mother agreed: "They really did enjoy it. That's the main thing" (Nell Lewis, October 5, 2009). Clearly, just like with the fishing, economics became second to giving enjoyment through preserving and sharing resources. This narrative contains several typical aspects of a story about the Lewis family: not only is the do-it-yourself theme there, but, while interested in the prospect of having a "sideline," the family considered the more important goal to be enjoyment and giving others enjoyment through sharing. When I reminded Muriel Meserve about the Girl Scout ceremonies—particularly ceremonies that use the bridge between the island and the mainland to symbolize a girl bridging from one Girl Scout level to the next—she went on to describe, with a laugh, other secular activities sponsored by community groups:

> Almost every group at one point or another has done something over there. It just seems like *the* place to do it. Used to have fireworks there, and then sometimes they would do them on, you know, on the side and they'd *watch* 'em here ... until that got to be, you know, reasonably unsafe. It was a given on Fourth of July that you were going to be one place or the other with the fireworks. (October 5, 2009)

In 2010, the merchants of Lambertville and New Hope not only reinstated the Fourth of July fireworks but also added fireworks on other Friday nights through the late spring and summer. The Lewis family responded by again stringing Christmas lights along the footbridge and the abutment. They then drag resin chairs to the island's point where the view is best and post a family member on the mainland side of the bridge to welcome visitors onto the island to see the fireworks.

The name change from Holcombe Island to Lewis Island not only reflects Bill Lewis's purchase of the island but also salutes his family's generosity, a quality this family takes as a collective purpose, and when they have not been able to fill the mandate as they would like, they are deeply troubled. The extended Lewis family experienced a profound impact on their family identity when the damaging floods of 2004, 2005, and 2006 compelled them to close the island repeatedly for safety and security reasons, and it took several years before they could fully reopen it and follow the family mandate to share the island with the community. Specifically, sharp objects redistributed from upriver and uneven ground created by erosion posed danger to people, such that there was also financial risk great enough to outweigh the "private property, travel at your own risk" sign. Financial risk also stemmed from the fact that the two buildings on the island, which lost doors and walls, were especially vulnerable. Cleanup was slowed significantly, because the flood downed the only utility pole that brought electricity to the island.

Significantly, the closing of the island due to flooding roughly coincided with Fred Lewis's death, and this created anxiety among townspeople and family members, who feared that changes could in some way become permanent. The family members did not completely agree on how long the gates they had built on the mainland side of the bridge should stay closed, but all were clearly unhappy doing so, for they felt that they were not fulfilling Bill Lewis Sr.'s mandate to "share the island," and thus their actions were in conflict with this Big Story, no matter how much they felt public safety or financial risk warranted the temporarily closed gates. Their chagrin at having the island closed was mirrored by dozens of dog walkers, swimmers, fishery visitors, and others who like to come to Lewis Island from time to time to connect with the river. Pam Baker describes how they were able to reopen the island by steps in 2009:

> It's starting back up again. Yeah, we're starting to, I mean, this year, we definitely used it more than last year, and I'm glad, and when we were young it was always [being used] that way. I mean, because as soon as fishing season started until [Fred and Nell Lewis] moved out in October, we were just always over there. And we're getting, oh, I don't think it'll ever be that busy again, just because we have [other responsibilities], and we can't get her [Nell Lewis] over there 100 percent, but it is being used more, and I think one of the nicest things we heard about Mur's [birthday] party was it was so nice to see it lit up. And that's kind of what

we—[Charlie Groth: it's been a long time]. It has been, well, and it's been dark in lots of ways for five years, since Grampy's death. (October 4, 2009)

Muriel's seventieth birthday party, held just two months before this interview, was transformative for the family, because it was the first large family event they'd been able to hold on the island since the flood cleanup, a period of literal (no electricity) and figurative darkness on the island. Pam points out how the practical situation caused by flood damage entwines with Fred Lewis's passing and absence from family leadership, passing the stewardship of the island through the various floods to the next generations.

It is important to keep in mind that the family's and community's discomfort with the island's closing may be part of a larger cultural experience. In *Mapping the Invisible Landscape: Folklore, Writing, and Sense of Place*, Kent Ryden asserts that "sense of identity may be one of the strongest of the feelings with which we regard places: when our meaningful places are threatened, we feel threatened as well" (1993, 40). In this situation, the flooding threatened the family's identity as sharers of the island, but also threatened the island itself through erosion, and threatened the access of many people who enacted their identification with the river through their presence on Lewis Island.

Sharing the river with contemplative visitors has special meaning for the family. These visitors' presence reflects the island's spiritual side, which is composed of late twentieth-century meditation practices, earth-based folk spirituality if not more formal neopagan or Native American spirituality, and more mainstream Christian (mainly Protestant) religion. The island has hosted a good many weddings and baptisms on its banks, and during the summer, Centenary Methodist Church of Lambertville regularly holds Sunday services as a way to beat the heat and live the popular Methodist hymn, "Shall We Gather at the River?" Informally, visitors would "walk out their troubles," as Fred and Nell Lewis used to say, and the couple would frequently witness people praying, crying, or meditating. Muriel Meserve concurs: "I can't tell you how many people have told me stories about having huge problems and watching the water and having, you know, a religious experience with it, having it all come together and understanding what to do, and being totally calmed down there.... The island has that effect on people" (November 7, 2003).

The Methodist hymn "In the Garden" also resonates for people making spiritual visits to Lewis Island, for many take a walk to Kathy Berg's garden, up past the crew's cabin and the Lewises' house. Calling attention to her grandparents' unusual generosity in allowing their neighbor Kathy to clear land and plant a garden on their property, Pam Baker says,

> Here again, they're like, "sure, go ahead. You can do that. It's fine," where ... there's not many people who do that anymore, who say, you know, this is my land, but you can go ahead and have a vegetable garden on it [laughs]. . . . You know, I'm not using that part, and she makes these beautiful things, and we ... have this great relationship come out of it. Again, that's just another case of not [being] afraid to share. (October 4, 2009)

In a pay-it-forward gesture, the garden's existence tells the Big Story of sharing the island with the community, Kathy Berg then allowing other visitors into her garden. When Berg died in 2016, members of the Berg and Lewis families, along with other local funeral attendees, began collaborating at the memorial reception to ensure that her garden would be tended the following growing season to keep the community resource going.

A trip to Lewis Island is for many a religious or secular spiritual journey—a "liminoid" ritual, anthropologist Victor Turner (1974) might say. Although some people are prompted by newspaper articles to drop by the fishery to celebrate the arrival of spring, many people independently plan to visit the Lewis Fishery. Some, particularly those from the immediate area, may visit frequently, with or without plans to buy shad. Others may visit just once during the season, telling crew and family members that they are making their annual pilgrimage to Lewis Island to see the nets pulled in. Individuals, families with small children, and dog walkers all integrate their sense of the spring season with their sense of place. This widespread association of spring rebirth and Lewis Island imbues the place with a cosmological, almost numinous dimension, which is consistent with its identity as a Protestant place of worship (see chapter 3 for a detailed discussion of the relationship between Fred Lewis's faith and the island).

On the other hand, the ambivalent push and pull between water and island, river and people, seems much older and perhaps even intimidating, like some depictions of the Old Testament God and nature-based spiritual forces. While the fishery family will act not only to protect the river but also to protect the island from the river's erosion, they are quick to point out that one can never forget the river's power, which has the potential to destroy a home or a life. The island's kids learn early to let a swift river take a beach ball rather than risking life to retrieve it. Muriel points out that the river, fluid as it is, is so powerful that it can even change the shape of an island:

> Dad always said, you know, the river takes care of its own. It brings and it takes away. And he would not have been nearly as upset about this erosion [from the 2009 flood] as we were, I don't think, because it was just taking it down to what it was before. (October 5, 2009)

Here, she notes the variety of reactions stewards might have to their conflicting roles of protecting the river and protecting *from* the river, both of which are tied up in a wise person's acceptance of the river's nature. When people hear of the way the footbridge is built to break in floods and be rebuilt afterward, they often try to think up ways to avoid that work. The family and crew learn to patiently explain the danger in creating a barrier on the creek that could lead to worse destruction of local homes. Over time, river people give up mastering the waterway and seek symbiosis instead.

Just as the Delaware River reaches the spiritual height of being "always," Lewis Island also reaches a mythical height of being a sort of "everyplace." The extended Lewis family has many stories of visitors, and while some form ongoing relationships with the family, others appear more as anecdotes depicting larger themes of humanity. One particular story Fred Lewis told, whose variations have been told by other family members, too, is not clearly rooted in either fact or fiction. Sometimes it is set on Lewis Island with Fred as a main character, and sometimes it is told in the third person about no one and nowhere in particular. In the more specific version, the story begins with a nameless man who comes to Lambertville from an unnamed place, stops by Lewis Island to enjoy the view, and strikes up a conversation with Fred Lewis. "What are the people like here?" "Well," responds Fred, "what are they like where you come from?" "Oh, they're horrible there—they lie and cheat, unfriendly and won't help anyone but themselves." "Well, they're pretty much the same here," Fred responds and lets the man go on his way. Another time, another person visits Lewis Island and asks Fred the same question: "What are people like here?" "Well," Fred responds again, "what are they like where you come from?" "Oh, they're wonderful! They are so honest, and kind, and friendly, and they'd give you the shirts off their backs." "Well," concludes Fred, "they're pretty much the same way here." When Fred told this story to me, he would usually end it with a laugh, not offering any concrete details to lend the story credence.

Because this story works on the level of myth, there is little need to determine whether it "actually" happened or not. In her examination of flood stories in Fort Wayne, Indiana, Barbara Johnstone asserts that "while personal experience stories, rooted in places, tie speech communities together, placeless myths bind larger groups together—ultimately, perhaps, all humanity" (1990, 127). With a mottled identity as both personal experience narrative and myth, this story forges both types of connections. It may be that the family told the story enough times that eventually it became associated with Fred, and it may even be that Fred heard the story, then reenacted it at some point during his decades of living on the island, prompted by the common question, "What's it like around here?" When the story is not set specifically on

Lewis Island, it may still be linked to the island with phrases like, "It's like the story about the guy who ...," "it" being the nature of Lewis Island and its effect on people. Either way of telling the story elevates the island to mythical place and its visitors' stories to stories of the human condition. Muriel Meserve agrees that the island is a special place: "And, and that's the thing, I think there's ... a little bit of a lot of things over there, so it's whatever you're looking for, you can pretty much find it" (October 5, 2009). Although the story does seem to say that what you see is what you get, it is also a story that stewards the local community. In telling and enacting the Big Stories of tradition, environment, and sharing the island, Lewis Island people tip the scales such that visitors experience civility, commitment, and caring. This kindness then encourages visitors, led by Fred Lewis, to do their part to maintain civility in the community.

THREE

The Captains: Between Myth and Legend, Article and Anecdote

The free bridge between Lambertville and New Hope is held up in the middle by five pylons made of stone and mortar, tapering in such a way that they may seem slender from a distance, but they can carry tons of bridge and car weight while being pressed by waters several feet above flood stage. During a haul, the captain looks downriver, then lines himself up with the nearest pylon to gauge how far out the net and boat have traveled—we're almost to the fourth pier, he might say. Then, based on that measurement, the height of the river, the strength of the wind, the amount of net being used that haul, recent catches, and the number of customers waiting on the bank, he decides where to turn the boat, when to ease off the oars or row faster, whether to instruct the landsman to hang back or walk slowly down the shore, and when to pull the pin holding the last bit of net and row like crazy. In a way, the bridge pylons resemble the fishery captains themselves, wading out in the water, each strong enough to withstand a pummeling by the elements and still stubbornly stay in place. For each captain, there is also a body of character anecdotes which, like stone and mortar, shape the man and serve as a guide for others. These "others," who become aware of the fishery at different points in their own lives as well as different points in the fishery's history, tend to embrace one captain or the other as the fishery's chief "protagonist." Character anecdotes about the captains are an important part of the narrative stewardship system, because they link the Big Stories to individuals in discrete episodes, making the people larger than life at the same time that they make the qualities accessible to others through the captains' humanity.

Within the system of narrative stewardship, character anecdotes about the captains ride the border between the Big Stories discussed in chapter 2 and the microlegends to be discussed in chapter 4. Of course, types work better as description than as proscription: stories muddy the typologies as quickly as they suggest them. In his wonderful ethnography *Storytelling on the Northern Irish Border* (2008), Ray Cashman notes that in conversations whose subjects span the community's living memory, stories of local figures can be classified as anecdote or legend depending on the relative time-placement of the subjects and audience members:

> The anecdote memorializes contemporary persons or ones within living memory, whereas the personal legend memorializes those who died before the tellers and listeners were born or cognizant of a given personal legend's protagonist. This temporal dimension is relative, hence one reason for overlap between the two genres. One person's experience with a particular character may be narrated in a first-person narrative as an anecdote, and the same narrated events may be recalled as a secondhand personal legend by others in a later generation. (Cashman 2008, 109)

These generations may be generations of the Lewis family, crew, Lambertvillians, or fishery customers. One conversation on the riverbank may span three generations in any one of these categories, including visitors who recall Bill Lewis Sr. personally, family members born after his death but with considerable direct experience of Fred Lewis, and crew members who joined the crew since Fred Lewis's death in 2004. Moreover, legends—some even with myth-like stature—grow from personal experience narratives, because we're talking about "real" life, twentieth-century life, give or take a few decades, and therefore have not yet gathered much in the way of "the mists of time." Some of these anecdotes even come to audiences through the voices of contemporary reporters, taking on the mantle of objectivity and modernity, even as they build the Big Stories, encompassing both fact and symbol.

In a sense, the relationship between character anecdotes about the captains and what the fishery itself stands for is a simple example of synecdoche, the type of figurative speech wherein a part stands for the whole. Nevertheless, a brief description of the captain's role shows *how* his nature affects the nature of this social structure. The word "captain" seems a bit outsize, because today "captain" usually refers to the nautical leaders of large ships: either captains bearing the enormous literary weight of Ahab and Hook or contemporary naval captains who operate more like the CEOs of floating corporations. Here, though, it's important to remember that Lewis Fishery captains are more the captain of the fishing crew than they are the captain of the boat. As such, their heritage of male leadership of small groups called "crews" bears resemblance to crews of mummers in Northern Ireland (Glassie 1983) or sports teams, where camaraderie intermixes with formal deference to the captain, who takes on the decision-making, organizational, and representational tasks of the body.

The captain's place at the top of the fishery hierarchy is practical in each haul as crew members ask him for work assignments and carry them out. The captain's role becomes obvious when you hear the children and captains' wives talk about the men they interact with quite differently in other contexts. Although generally united with her husband in one breath as "Fredn-Nell," when it comes to talking about the fishery, Nell Lewis would usually take a step back and either not take part in the conversation or speak a little more formally about Fred. This situation is special for Sue and Steve Meserve, for—unlike Fred and Nell, who lived in a time when men's and women's roles were more separate—Steve and Sue occupy overlapping spheres in relation to construction and maintenance; Steve follows a long family tradition of DIY, but Sue is a professional carpenter like Bill Lewis Sr. However, having headed a shop (again, a "crew" of men), Sue understands the masculine hierarchy and holds her tongue on the island in front of the crew. Rather, she and Steve have worked out a partnership whereby they consult with one another without an audience, then Sue is in charge of some projects and Steve directs others himself.

The captain's authoritative position carries over to narrative stewardship roles with respect to working with the media. As discussed in chapter 2, Renée Kiriluk-Hill of the *Hunterdon Democrat* went to Fred Lewis for perennial article quotations for decades, then transitioned to consulting with Steve after Fred's death. The captain's authority can be seen in relation to the relative hierarchy of media sources as well, when reporters and photographers interact with more than one family or crew member. Pam Baker says that when requests for quotations come to the family, "[the] *New York Times* goes to Steve. I can talk to Renée; that's about it" (October 4, 2009). Here, Pam reflects not only on the prestige of the *New York Times* and the intimacy with the county reporter but also on her place in the family and crew with respect to authority at the fishery: the captain is the ultimate representative.

The captains' characters are elucidated through the character anecdotes and reflect communal values intrinsic to the stewardship functions of the fishery and the family. Ray Cashman defines the "local character anecdote as a reportedly factual, brief, first- or third-person narrative about known or knowable local individuals" and breaks down the category into two types, "contemplative and comic," which can overlap (2008, 96). Cashman's analysis of character anecdotes told at Northern Irish wakes and *ceilis* emphasizes *characters* in the sense of someone being "quite a character" or "a local character," a usage that finds its way into storytelling at the Lewis Fishery. In relation to the fishery captains, my usage points toward "having good character" but also refers to stories that outline the captains' idiosyncrasies and foibles, making them human, accessible, and even quirky. Captain anecdotes characterize not only the individual men but also the fishery and the Lewis family; in fact, character traits of the captains may be shared with other members, both males and females. After covering character anecdotes that exemplify traits such as being stubborn, exacting, generous, firm, steadfast, kind, or gentle, I will discuss differences between the captains as *figures*, showing how the passage between captains matches larger themes in the fishery's history and ethos.

For the fishery family, the root of the environmental Big Story always lies in anecdotes about Bill Lewis Sr. (also called Cap'n Bill or Poppy), who, along with his brother Theodore (Uncle Dory), was known for the character traits of intelligence, ingenuity, and perseverance—the last trait often framed as "stubbornness." Muriel Meserve highlights this last quality when she asserts, "They say my grandfather was stubborn, only outdone, I guess, by his brother, Theodore, the one that built the [paths on the island to support the fishing]—I mean, they would argue black is white, and never the twain would m—; I mean, neither one would give an inch" (October 5, 2009). Her mother agrees as Muriel continues: Uncle Dory "would, you know, get something in his head and you weren't going to change it. Poppy was the same way. [Nell Lewis: Yeah.]" (October 5, 2009). Here, the conventional start to Muriel's story—"they say"—also adds weight to her assertions; there are enough people who would agree that these people become anonymous and uncountable as the vague term "they" suggests. On the other hand, Nell's simple "yeah" corroborates Muriel's view of their stubbornness by adding eyewitness support from a specific source connecting the present audience with the two men who died decades earlier.

While clearly the stubbornness could be frustrating for the rest of the family at times, the next four generations of the Lewis family also express appreciation for this quality, which contributed to saving the fish population. When Pam Meserve Baker describes her great-grandfather's character trait, she does not remember the term "stubbornness," foregrounding "perseverance" instead:

> But he just, and he *did*, he just . . . pursued. You don't see that kind of perseverance a lot, so this must have been, you know, years and years of letter writing, and then to go and testify, and that's a great legacy for the kids to see. I mean, this was back when, you know, people just didn't do that, you know, it's like *Mr. Smith Goes to Washington*, OK [switching to Poppy's perspective but using her own expressions] . . . "because you have to"; you know, "you're screwing *everything* up." [Switching back to her own perspective] And they *did* clean it up; it took *years*,

though, and that's the thing. It's like, this didn't happen overnight. This is perseverance.... Growing up with it, it was just ... as a child, it doesn't mean anything until it becomes *your* history, and then at some point, whatever it is, when you have children, when you start,... you have that realization of all this history ... it impacts *you*, that you're *part* of the history. It's like, oh wow, that's like, that's pretty incredible. (October 4, 2009)

Like Steve, Pam links the character trait to actions that led to a broader trend taken up by others. She invokes the weight of far-reaching history by mentioning both the political act and its historical record as well as a family history that will continue into future generations and can be owned.

Access to history also takes the form of the character trait being positive as well as negative, as in the same interview Pam called the trait "orneriness," which her mother laughingly labels "stubbornness," "a *huge* Lewis trait" (October 5, 2009). Muriel credits the fact that her grandfather, Bill Lewis Sr., ran a "tighter ship" than the other fisheries as a key to keeping the crew going all these years. "But, half of it too was Lewis stubbornness. I mean, he just wouldn't quit! When the rest of them gave up with no fish, he just kept, he was out there every year, which is why his [license] never ran out [Nell Lewis: yeah], because he *did* it every year. And it was just sheer stubbornness [laughs]" (October 5, 2009). In my first interview with Steve Meserve, he linked Bill Lewis Sr., stubbornness, and the environmental Big Story this way:

My grandfather's father just—because, I guess, he was stubborn enough, kept at it and left us what we've got now. And, I think, if he hadn't done what he did, you probably wouldn't see any shad in the river. I think, because he was out there all the time showing you that, no, there are no fish, because of that, the pollution is so bad these fish won't survive. And if these fish can't survive, what else can't? (April 22, 1996).

Here, Steve connects persistence of the individual with the group's persistence and links its positive impact to the health of the river, the shad, and back to the river's health again. It is important to note that Steve builds the import of his great-grandfather's contribution of persistence by also connecting that "stubbornness" with intelligence. Toward the end of the statement here, it is unclear whether Steve is repeating his great-grandfather's trajectory of thought or furthering his own inquiry into the present, thus subtly using narrative to entwine five generations' commitment to stewardship.

Steve may resemble his great-grandfather in his persistent commitment to the fish and river through the fishery's activities, as well as in his unassuming manner, which Nell Lewis emphasizes as she tells one of the core anecdotes that promotes the environmental Big Story. Her story emphasizes the drama of Bill Lewis Sr.'s testimony situation as she highlights her father-in-law's courage and skills, as well as the class-based underdog aspect of the incident:

> Well ... the fish got more scarce all the time, and then the river was dirty, and from down in Philly, and Fred's father and some other people ... who went down. And he [Bill Lewis Sr.] said that they went in to have this meeting, and some of these commissioners who were down in Philly,[1] they thought, "that old hick." He's a skinny, tall skinny man. He looked like an old farmer, you know. And they thought, "he doesn't know," you know, "he doesn't know much, and we can put him in a tailspin." Boy, he got out [chuckles] the book of the number of shad, how much they caught, when they caught them and all. He had all of that ... how the river was at that one day, you know, every day, just like Fred used to do it, like Steve. ... And, uh ... so, they said Poppy really gave them everything they wanted to know and they said he really put them in their place. (October 5, 2009)

Nell Lewis goes on to paint a picture of Bill Lewis Sr. as a quiet man, thus showing the courage of his act, confirming her understanding with her daughter, Muriel Meserve, and myself.

> NELL LEWIS: He wasn't a person to get up and talk, really, was he?
> MURIEL MESERVE: No, but he could ...
> NELL LEWIS: I mean, Fred ... could get up and talk, but Poppy, I mean, he would talk to people along the shore, if somebody asked him about the shad or something like that, but he's not one to get up in front of people, but he did that day. He knew what they were doing, ah ...
> CHARLIE GROTH: And he knew what they needed to [Muriel Meserve: know] know, and do. Nell Lewis: Yeah.
> MURIEL MESERVE: And he was persistent enough with it that he got through not only to two states, but to Washington, finally, about what needed to be done out here. (October 5, 2009)

Again, this account emphasizes the persistence of making the hauls, recording the data, and telling the story of the alarming drop in shad population. However, it also adds the courage it took for Bill Lewis, in particular, to put himself out there, when he was much more comfortable in the background. Indeed, in the 1937 photo series of the crew, Dory Lewis seems to pose for the camera, while his brother, the captain, looks down and seems unaware that he is the subject. Nell Lewis also hits on another thematic type: the story in which the local outwits seemingly more sophisticated outsiders who underestimate his or her intellectual abilities (see Cashman 2008, 182–85, for a discussion of this story type found by Henry Glassie and himself in Northern Ireland and found by Pat Mullen in Texas). This ubiquitous dynamic appears between American cowboys and city slickers, and between Scottish clansmen and English rulers. It appears in gendered and racialized contexts, such as the movie *Fried Green Tomatoes* (Avnet 1991), in which local women and black men outwit a white male sheriff to protect one of their own. Looking like a "hick" and not expected to perform well, Bill Lewis Sr. performs well in the politicians' own arena—paperwork in the form of ledgers. He wins their attention and an outlay of resources to benefit a rural region usually forgotten.

Nell Lewis's story of the "hick" who "put them in their place" hits on a second character trait that runs through the captain anecdotes: native intelligence. Mentions of Bill Lewis Sr. usually start with a reference to the young age at which he began not only fishing but *running* fisheries, evidencing his know-how of shad, the river, and people. This intelligence covers both work smarts and science through involvement with homegrown and scholarly projects. Fred Lewis recalls his father building a glass-bottomed box, which he held over the side of boat and looked through to observe the behavior of the shad as they encountered the net, then used this information to fish more successfully. Specifically, he found that trapped shad would follow the net downstream to find an opening, and then circle back when they gave up. With each pass, the fish would go a little farther, until by the third pass they would usually clear the net. Cap'n Bill concluded that he needed to manipulate the net through rowing, poling, and directing the landsman's actions such that the sea end of the net was back to shallow waters before the third pass. This interest in observation and experiment continues. In addition, the fishery captains have collaborated through the decades with academic and government researchers, putting the crew, tools, and process at the disposal of science in tagging, sampling, and record-keeping projects.

Bill Lewis Sr.'s family points to his record keeping as intelligence of foresight: it was the data in the ledgers that convinced the government officials to do something about the shad problem. This foresight intertwines with an almost Ahabian drive—but one in which the captain saves the silver fish rather than killing the white whale. Due to the man's efforts and faith, the sons and great-grandson and their crews faithfully continue the traditional work and confirm the father's certitude. Muriel Meserve also refers to her grandfather's intelligent foresight when she says, "He kept saying, 'if you clean it up, they'll come back,' and of course they did" (October 5, 2009). The fish resurge, and although the father passes away, he becomes legendary in his important role, and his legacy surges through the next half century of family history in the form of the fishery's environmental Big Story and through his mandate that his children and grandchildren share the island.

The island's built environment also evidences the captains' intelligence through their abilities to build or fix anything—a boat, a bridge, a house, an eroding island, what-have-you. The stories about building the house on the island at once show Bill Lewis Sr. passing his can-do attitude, knowledge, and work ethic to his sons, Bill Jr. and Fred. Fred Lewis recalls, "In 1933, after I got out of high school . . . it was in the Depression, and my brother and I had fished that season. . . . We got finished with the fishing around the tenth of June and Dad says, well, you boys don't have a job; you may as well build a house. So, we started building the house on the island then. And by October we could move in" (March 1, 1996). As Nell Lewis and Muriel Meserve narrate the house building, the process was not quite as simple as Fred's telling makes it sound. Nell explains:

> [Fred] and Bill did a lot of the—quite a bit of the work on it, eh, as Poppy laid out the plans and all. And he [Poppy] was working three days a week down at the Union Mill, and so when, those days, why, Fred and Bill [Jr.] would do some work, and [Fred] said they put the floorboards down. And . . . they [had] them all real nice and snug and had 'em all ready. Poppy came home, he said, "You're going to tear them all up!" "Why?" [laughs] And here, they were, they put 'em too snug. He says if the river came up, that would buckle the floors. (October 5, 2009)

Muriel then gives the coda: "And obviously he knew what he was talking about, because those floors are still fine after all these floods [laughs]. But, yeah, he'd come home and he'd inspect what they did, and if it wasn't right, then they'd do it again! [laughs]" (October 5, 2009). This story shows Fred and Bill's carpentry abilities, for they were able to make the floor snug. Bill Sr.'s carpentry skills appear in his understanding of wet wood, but the story also shows his knowledge of the river's behavior, an intelligence son Fred and great-grandson Steve show to crew, scientists, and the press through their tenures as captain. In this story, Nell and Muriel also introduce the character trait of determination to meet high standards, something that can be related to persistence and stubbornness, but that also has a moral element.

Related to the Big Story of civility, meeting higher standards of behavior mainly takes the form of "doing the right thing," not tolerating fighting or excessive drinking. Not pointless perfectionism, holding the bar high here sits in the relaxed context of a fishery where mud, flood, and shad slime challenge pretensions of propriety. Ripping up the floorboards, for example, was not because the floor was not perfect; it was in fact too perfect to withstand the power of the river. Similarly, Fred Lewis remembers his brother and himself constantly "fighting" their way through the two years between Bill Sr.'s and Bill Jr.'s deaths, when the two brothers captained the fishery together. This, however, was not the brawling of drunken crew members. Rather, the fighting involved disagreements between two headstrong brothers who had collaborated on fishing the Malta Fishery (a two-person operation) but had not been trained in captaining the larger fishery and its crew. Repeating the history of their father and Uncle Dory wrangling before them, Fred and Bill Jr. might stubbornly fight, but eventually worked it out in the family business. When it came time to pass on his leadership role, however, Fred learned from his early experience as captain and more actively taught his grandson the role.

Fred also chose a different path from his father with respect to educating his children, as suggested in an anecdote that also exemplifies the Lewis stubbornness and high-mindedness. Muriel Meserve claims that "the only mistake [Bill Lewis Sr.] ever admitted making was not . . . letting Dad go to college" (October 5, 2009). The decision came amid the Depression, in the wake of Fred's elder siblings' lack of success, and in relation to a disagreement about football. Bill Lewis Sr. thought football to be extremely and unnecessarily dangerous ever since Bill Lewis Jr. broke his neck on the town team with the lasting effect of blacking out every time he turned his head "just the wrong way" (Nell Lewis, October 5, 2009). Nell tells the story here: "Well, [Poppy] said to Fred, you know, 'I'll send you, if you promise not to play football,' and Fred said, 'I can't promise it [Muriel:—which is also the Lewis stubbornness] [Muriel laughs] 'cause I'm going to play football.' He was football crazy. And so that was it. And of course he was the one [of his siblings] that would've gone [to college]" (October 5, 2009). Although he never did go to college, Fred Lewis had a reputation for being "learned" and "smart," as Kiriluk-Hill refers to both Fred and grandson Steve: "The thing you learned when you dealt with Fred was that he not only was dedicated, but *very* learned about not only shad, but the river, the community, the environment, which was a treat, because it wasn't the same old story time after time" (June 3, 2011). Knowing lots about shad characterizes the fishery captains, but this knowledge doesn't stop with fish. Fred educated himself through experience, correspondence courses, and reading magazines such as *National Geographic* and *Scientific American*. Fred's reaction to his father's decision perhaps explains how formal education moved into a more central place in the family values: Fred was determined to send his own children to college in a day when higher education for girls was not a foregone conclusion, and his daughter Muriel says, "There was never any question that we were all going to go to college; I mean, that was just a given from Day 1" (October 5, 2009). Yet, it is important to note here that this is not just a story about education and stubbornness. The anecdote tells of two people dedicated to doing the right thing when the right things clashed: protecting one's son on the one hand and refusing to lie to one's father on the other.

Fred's unwillingness to make a promise he knew he would break relates to two other character anecdotes about Fred Lewis's resolute voice. In one story, some local leaders wanted to reestablish the Lambertville–New Hope fireworks after the town of New Hope had stopped shooting them from New Hope and after Fred Lewis had tried staging the fireworks on the island and decided that it was not safe there, either. At one meeting, the organizers urged the mayor of New Hope to approach Fred again, thinking that the mayor's long history with Fred would hold some sway. Nell Lewis tells the story here: "And [Fred] said no because [of] the danger. And at the time, Jim McGill was mayor of New Hope, and he said they asked Jim to ask you because we knew Jim. We've known him ever since school days [laughs], you know. Yeah, he was older than us, by, oh, I don't know, maybe about four years, but he said, 'I know Fred Lewis better: and if he says no, it's no.' And so, he never come over to ask for it" (October 5, 2009). The use of "ask *you*" (emphasis mine) suggests that Nell is going back and forth between her recollection and re-creation of the story. Here, she perhaps plays the part of whoever first told Nell and Fred about the incident, and further along she clearly speaks Jim McGill's "line," which is the climax of the story. Nell underlines the mayor's point about Fred's steadfast word by corroborating that the mayor never did bother to push the point with Fred.

Fred Lewis was to be taken at his word, and in another comic anecdote Fred told, people's great trust in him has humorous results. For several years Fred served on the Lambertville school board and was part of the team that built the new school in 1968, the one that Lambertville children still attend today. During his stint on the board, Muriel applied to work as a teacher, having returned to Lambertville after college with a husband, her small son Steve, and a teaching certificate. When it came time to approve her hire, Fred abstained from the vote as was only proper—and so did another board member, much to Fred's surprise and chagrin. At the close of the meeting, Fred asked his co-board member, "Why did you abstain on that vote?" and received the answer, "Well, I saw that you abstained, so I figured there must be something wrong with her!" Fred replied, "Well, I abstained because she's my daughter!" In this story, Fred's integrity is established through his choice to abstain from the vote, thus refusing to use his position serving the community to advance the interests of his own family. The public's awareness of his integrity and intelligence is evidenced in the other board member's statement, which at once establishes the man as something of a newcomer—new enough not to know that Muriel Meserve was Muriel Lewis, a local girl—but nevertheless someone well enough acquainted with Fred's character to base his own vote on Fred's judgment. The ironic humor of this story becomes even more evident in retrospect, since Muriel Meserve turned out to be one of the most respected and well-loved teachers in the school's history, a model of competence and good character in her own right. In Lambertville, Muriel Meserve is known as one of the proverbial pillars of the Methodist Church, often aided by her daughter Pam, and both fulfill the role of the faithful Christian modeled for them by Fred and Nell Lewis.

A review of the captain figures would be incomplete without discussion of Fred Lewis as a man of faith who is at times larger than life, almost a spiritual figure himself. In this way, Fred is inseparable from Lewis Island, which is for many a spiritual site. On the one hand, the numinous quality of the place draws together the Methodist traditions, the space, and the people in it. Pam Meserve Baker tells the story of a special spiritual experience she had that connected her to the river, to her grandfather, and to their Christian tradition. As a girl, Pam did not feel that becoming a member of the crew was her lot and became involved in other community and family activities more typical of females. However, one spring, summer, and fall in young adulthood she spent considerable time and muscle—"sweat equity" she labels it—to help her grandfather haul rocks around the island to avert erosion. The following Good Friday, her grandfather invited her to row the boat with him, something he had never before done. As she tells the story of the conversation that night, with her asking questions and her grandfather answering them, she makes a "breakthrough into performance" (Hymes 1975) and portrays the scene as a spiritual one:

> For whatever reason, on one Good Friday, he let me row. I don't know why. And it must have been after my father died. And I don't know why I was here. You know, it's one of those—because it must have been before I got married . . . but, the two of us rowing . . . out, you know? And that's my memory. You know, when I'm ninety-three and I've got Alzheimer's or whatever, that's what I'm gonna remember. And . . . the *calm* of it; "it's the most beautiful place in the world. Isn't it?" "Yes, it is." "The sunset on the water, isn't it the most beautiful?" "Yes, it is." I mean, it, it was a perfect night. And, as I say, because it was Good Friday, and . . . that's my perfect. . . . Because he *was*, he . . . he just walked and lived it. (October 4, 2015)

The "it" she names here is Fred's faith, anchored in the story of Easter as well as the relationship with the river. Moreover, in the context of the way both Pam and Fred have spoken of the river and the island over the years, it would not be an exaggeration to interpret their comments about the evening on the river as comments about their faith, with their duty to care for "the most beautiful place on earth" as a parallel to ministry within the church (which, remember, holds services on the island in summer), which in turn follows the model of Jesus's relationship with people. She specifically places them within the haul as going "out," venturing on the hardest work and also the peaceful stage of the haul ahead of them. They were "rowing . . . out" together in their faithful venture as fishermen and Christians, two identities linked in Christian iconography and the stories of Jesus and the disciples, his "crew." Fred himself would connect the Christian stories with the fishery, for he was fond of telling of the miracle when Jesus fished with a crew: there were no holes in the net. Of course, this connection also bears a humorous contrast, for the fishery captains were and are forever mending the net.

Bordering on myth, Pam's story's exact time setting is hard for her to place against the more worldly outline or her life. The exchange she recalls also reflects the two personalities, Pam's talkative enthusiastic voice playing against "Grampy's" terse but gentle comments. This story of a gentle river seems to wed the caring relationship with Jesus with an almost transcendentalist spirituality, giving both the river and the island a numinous quality and offering yet another picture of Fred as a spiritual guide and example. Only partially in jest, Pam says that in her head she sometimes thinks of WWFD (what would Fred do?) as a version of WWJD (what would Jesus do?) to give her wisdom and guide her actions, whether moral or practical. This problem-solving technique almost turns the character anecdote inside out: she imagines what story might emerge in a situation based on the character traits of Fred, whom she casts in her imaginary story. The process attests to the power of character anecdote as guidance, and it also simultaneously shows Fred Lewis as both larger than life and accessible through relationship and humor. As the stories of stubbornness suggest, Fred Lewis was not a perfect man.

While at times represented as formidable about the island rules against drinking, swearing, and throwing stones, Fred is also shown in family stories to have a vulnerable, impish, even silly side. His enjoyment of good-natured "tricking" became apparent in my first interview, which happened before the season started in a winter when floods had washed out the bridge. Explaining to me when the season would begin, he said, laughing, "We'll probably get them all together, maybe . . . two weeks from Saturday. And we'll tell them we're going to fish and we'll trick them into doing the bridge." In reality, the crew members were told that bridge building was the first order of business, since the family's more "saintly" tendencies keep the devilish sense of humor in line.

From time to time, Fred's grandson Steve recalls coming across the bridge with his wife for one of many family picnics, to be confronted by a rather ridiculous image of his grandfather. On one occasion, the pronounced heat and humidity on the Delaware in summer required an "anything goes" attitude that resulted in a completely mismatched outfit, including shorts, which Fred rarely wore. Unaware of Sue and Steve coming up the walk, he was caught between two tasks of caring for his family: stamping mole tunnels flat while holding barbecue utensils straight up like little flagpoles. Steve recalls collapsing in laughter at the sight of his grandfather in the distance, doing what he and Sue have called ever since "the mole dance," a little Lollipop Guild number with a barbecue fork, sandals, and socks. As he tells the story, he usually draws attention to the irony of this vision in contrast with his grandfather's larger-than-life status as a man of faith, well-respected community member, and captain of the Lewis Fishery for more than half a century. Similarly, I have been amused by the knowledge of Fred's quirks surrounding his discomfort with killing fish, an ironic characteristic given his profession and the necessity of the fish dying between being caught and being eaten. His aversion seems related to two endearing fictions: when the fish died, he called it "going to sleep," and when he packed them belly-up in wet burlap to preserve them overnight, he would call it "putting them to bed." This gentle demeanor matches an activity missed on the island since Fred's death: his whistling. Unusual in that it often lacks a tune or phrasing, Fred's whistling has been reflected in the remembrances of family, friends, and journalists.

When taken together, the stories of Fred's "stubborn" high-mindedness, sense of fun, hard work, and enjoyment of picnics and joking exemplify a narrative phenomenon Cashman discusses: "When featuring in a cycle of anecdote, a person may be seen as an amalgamation of varied and even contradictory characteristics and traits" (2008, 205). In literary terms, Cashman discusses a process in which a character becomes rounder and flatter through choices tellers make about which stories to tell on which occasions in which combinations. When told in more intimate settings of tellers and audiences who knew Fred and know each other personally, tellers are more likely to offer stories about the different sides of Fred's personality. This may happen within one conversation in which Fred is the topic, or different conversations may emphasize different aspects of Fred with the knowledge that the group rounds the image and balances the representation through the amalgamation of storytelling sessions.

At the same time, less intimate storytelling situations or a particular storytelling occurrence may "flatten" the actually well-rounded person. Again, Cashman (2008) elucidates this phenomenon, discussing several causes other than the level of knowledge the audience and tellers have of the subject. As the storytelling genre employed moves along the spectrum from the personal experience anecdote to legend, the "characters" in the story become more typified and flat (Cashman 2008, 165), a change that at the fishery fluctuates with the storytelling context and purpose: is the situation family members remembering, a crew member introducing a visitor to the fishery, or the captain talking to the press? Because they are characters in a story, the captains' personalities may also be "flattened" through choices that emphasize a particular message for narrative stewardship. Stubbornness and intelligence, which influence the fishery's longevity, might be emphasized in the Big Stories about tradition or environment. Cashman points out that the passage of time and loss of detail to memory may also flatten characters in the anecdote (2008, 165), and this flattening trend certainly affects the figures of the Lewis Fishery captains, who over time become historical figures.

Local character anecdotes and personal experience narratives about the captains intertwine with the Big Stories and emerging legends (see chapter 4) into a system in which tellers' and audience members' understandings of the captains reflect their own place in narrative stewardship, local community, and collective memory. Within the family and to some few old-timers in town, William Lewis Sr. is the essential figure of the Lewis Fishery. The stories of his buying and developing the island underpin the "sharing the island" Big Story, while his involvement with studying the fish and alerting the officials anchors the ecological Big Story. Together, these Big Stories gird the stewardship roles of fishery, family, and island. However, Bill Lewis Sr.'s younger son, Fred Lewis, eclipsed the Lewis Fishery founder in the public eye when he took over the sole leadership of the operation after co-captaining with his brother, Bill Jr. Fred Lewis became the protagonist of the second half of the twentieth century, becoming more remarkable as the tradition became more unusual in relation to changing culture and more important in relation to the Shad Fest. In a survey conducted at the 2012 Shad Fest shad-hauling demonstrations, visitors correctly identified Fred as one of the captains more often than the three other captains who at different times embodied the fishery. Steve Meserve, the current captain, was identified correctly about half as often, and Bill Lewis Sr. was identified only slightly less often than Steve. Most people acquainted with the Lewis Fishery do not think of Bill Lewis Jr. as a captain, if they know of him at all, mainly because the time he led the fishery—as Fred's partner, not as sole captain, at that—was very short and more than half a century ago, and so it makes sense that surveys correctly identified him as a captain least often.

In the 1990s, the captaincy hit, then weathered, a sort of crisis point as the role passed from Fred Lewis to his grandson Steve Meserve. While the conventional father-to-son inheritance traditional to American culture operated on one occasion at the Lewis Fishery, as a cultural norm it presented a challenge when Fred Lewis became too ill to lead the everyday operations of the fishery. An octogenarian with cancer, he could sometimes stand on the bank and guide the crew, but he frequently needed to stay inside when strength or weather failed. By this time, both Fred's son, Cliff Lewis, and Fred's son-in-law, David Meserve (Muriel's husband), had died young, leaving no local men in that generation to take the reins from Fred. Although his daughter Sue Lewis Garczynski had broken the gender barrier in the 1980s, her career and marriage had taken her from the area. Thus, Muriel Lewis Meserve's son, Steve Meserve, then in his mid-thirties, was given the new role called "operator," leading the crew during each haul. Fred retained the role of "owner," teaching Steve how to be a captain through daily consultations. Steve did not assume the owner role, which passed to his grandmother and then five years later to his mother and aunt, but simply retains the operator and captain roles. Similar to the boats' usage and other temporary fixes, the solution of splitting the lead role into operator and owner has lasted a long time. Since Fred Lewis's death during the 2004 shad season, an event in local history with mythic proportions, Steve's identity as the crew's captain and figurehead has strengthened.

Stories of Captain Steve are just starting to emerge, but his joining the ranks of Lewis Fishery captains has had a profound effect on the mythologizing of those captains, for one more generation and a switch in surname have thrust Bill Sr. and Fred further into the past, giving audiences more people and more history to grasp. Other than with family and old-timers, Bill Lewis Sr. has flattened as a character in the foundation myth of the fishery. Perhaps this flattening functions as a way for the community to master the expansion of historical content, whose sheer size may try the competence of tellers and listeners, who themselves are a changing population. The mythologizing of fishery captain figures becomes especially clear in the *Beacon*'s fishery coverage, which, since Joe Hazen's retirement, became more focused on Shad Fests than on shad fishing, flattening the story of the fishery as well as the people in it. The paper's articles emphasize the tradition Big Story, and in this flattening, stories of captains become more legendary and sometimes even inaccurate. In a Shad Fest 2000 insert in the *Beacon*, the haul seine method acquires a little verbal hoarfrost with descriptions such as "the craft mastered by [Fred Lewis's] ancestors more than 100 years ago" (April 27, 2000); the "ancestor" here is Fred's own father. Similarly, the year after Joe Hazen's death, a description under "Shad hauling" in the festival's list of events reads thus: "Fred Lewis, New Jersey's only commercial shad fisherman, will give demonstrations of shad hauling on Saturday and Sunday at 2 pm. Fred uses the seine technique originally taught to early settlers by the Indians" (*Lambertville Beacon* 1999). Here again, Fred Lewis's role is mythologized to the point that not only is he linked to a somewhat romantic view of white/Native American relations, but it sounds like he's doing a job alone that usually six or more people do. The seemingly intimate stance (calling Fred Lewis by his first name, an unusual choice for a newspaper) seems even more of a creative stretch when one recalls that by this time, 1999, Fred Lewis's activity was mostly confined to observation and consultation, often from a distance. This stretch honors Fred's real leadership but also reflects the *Beacon*'s separation from the fishery and the community, and the paper eventually folded in 2015.

At the same time, other somewhat misleading narration attests to the mythologizing of Fred Lewis, when people confuse Fred with his father, Bill Lewis Sr., the fishery's founder. These slips in accuracy and time appear in what seem to be verbal slips, such as when a reporter or community member refers to Steve Meserve's "grandfather" (Fred) when "great-grandfather" (Bill Sr.) is the accurate term. Cashman (2008) puts such an occurrence in context of narrative theory when he discusses a similar situation that he sees at wakes in Northern Ireland. He notes that tellers sometimes credit a particular priest with successfully laying a curse on an outsider, although the action is usually attributed to a different priest in variants of the story. Cashman then suggests that "in terms of dramatis personae, typology often takes precedence over individual uniqueness or biographical fidelity as stories about actual individuals make a transition from anecdote to legend" (2008, 162). In other words, the character type of the local priest is more important than the individual identity. In Fred and Bill Lewis Sr.'s case, their type as important captains from the past—perhaps compounded by character anecdotes typing them both as strong, gentle, intelligent, stubborn Lewis men—takes precedence over their distinct identities and an accurate timeline of their activities. The speakers might or might not understand the difference between the two people, and might or might not realize they have made an error. Nevertheless, errors confusing Bill Lewis Sr. and Fred Lewis have become part of the collective narration of the fishery's history, ushering it from local experience to regional legend and myth.

The mythologizing of Fred Lewis and his identification with the fishery's Big Stories were perhaps never more apparent than during the week of his death and funeral in 2004. Steve Meserve recalls that even when he and his grandfather were simply lecturing at Raritan Valley Community College or visiting the Bethlehem Shad Festival,

> even at those times, to me at least, his stature as a larger-than-life figure was apparent. So . . . really with his passing, that kind of solidified it to me, because I actually had to take time off from work to deal with that event, not from a personal mourning kind of thing, almost as a public figure out there to do, you know, answer the questions and give the history and reflect on the life of this man. (October 3, 2009)

With Fred's death as a "larger-than-life figure," Steve was quickly thrust into the position of figurehead himself. Fred passed away on April eighteenth, during the shad fishing season, and his funeral, scheduled within a customary time frame, fell on the Saturday morning of the Shad Fest. By the time of the funeral, not only had the obituary run in local and regional papers, but additional articles covering Fred's death supplemented the usual Shad Fest coverage that mentioned the shad fishing tradition. The nature of the previous years' coverage led to an impression of a more abrupt change in leadership than was actually the case. By this time, Steve Meserve had led the crew as operator for the better part of a decade, gradually assuming more of the other leadership tasks as his knowledge and confidence grew and as Fred's declining health necessitated. This process was difficult for both, since "Mr. Shad" had been such an important part of Fred's identity for decades. Ownership, habit, and great respect induced family members and reporters to represent Fred as the key figure in stories, and since by this time reporters tended to phone for a quotation and just send the photographer on site, news stories such as the 1999 Shad Fest insert discussed above made it appear as though Fred were more directly involved than he actually was and Steve had less of a leadership role than he actually did. Thus, the journalism in the years leading up to Fred's death helped set the stage for a momentous telling of the Big Story of tradition, accurately reporting that the Lewis family's and Lewis Fishery's changes were important to the local and regional community, even if the way this story was told was somewhat mythologized.

When commenting on the fact that the 2004 Shad Fest coverage needed to be about Fred's death, Renée Kiriluk-Hill said:

> That year you couldn't ignore [Fred's death], as, from a local viewpoint—even if the visitors [to the Shad Fest] didn't know this, it wasn't a big thing to them, [to] the *community* Fred's passing was a big thing. And so obviously that's what we had to focus on that year [in festival coverage], not to make people not want to

come [to the festival], because, oh, it'll be a sad thing, but, no, it was *celebrating* his life. (June 3, 2011)

Fred Lewis's passing was so noteworthy that an article honoring his life's work appeared on the front page of the *Trenton Times* in addition to the article on the obituary page. Unfortunately, not all media professionals understood the community's feeling for the Lewis family and Fred Lewis. Of such public import was Fred's death that a photographer from the *Trentonian*, a popular but not well-respected newspaper, snuck into the choir loft and took a photo of Fred's casket for the Sunday paper. The following day, the story circulated through the community, at least a few people alerting the family to the photographer's indiscretion. As Pam Baker recalls the incident, it is clear that she considers the photographer to have behaved badly:

> It was just uncool. It was just bad, you know, just bad, bad. Because I was just, I was so [laughs] pissed. Oh, God, it was awful. I mean, you would think it was [Princess] Diana or something, c'mon. You know? You know, write whatever you want, but ... I mean, I can't control that, but taking pictures from the balcony? Come on now. That was ridiculous. And because it was Shad Fest. That was the other thing. It couldn't have helped when he died. I mean, it was very apropos, the whole thing, and, you know, and it was just wrong. I mean, it was just bad. (October 4, 2009)

Five years later her anger still tongue-ties, and she uses terms that in their simplicity and repetition expose her outrage. While understanding that Fred's life and death were a matter of public interest, the community saw the action as crossing the line between the sacred and profane and coordinated their efforts to save Nell Lewis the shock of opening a Sunday paper and seeing her private grief published far and wide.

At Fred's death, the life of the man and the life of the tradition naturally intertwined, and the Big Story was told not only in the newspapers, both appropriately and inappropriately, but also through rituals related to his funeral, the festival haul seine demonstrations, and the crew's activities related to both. The crew served as pallbearers. The nature of the pallbearer team reflected their fishing work, including the fact that two women, Sue Meserve and myself, were included, but deferred to the men's traditional central position, falling back when fewer hands were needed. The service itself included three ministers, representing Fred's long relationships with the New Hope Methodist and Lambertville Methodist congregations. The music reflected the family's relationship with the river, as "Shall We Gather at the River," Fred's favorite hymn, was sung, thus bringing the physical space of the island into the sanctuary, where the crew and family now sat in the pews. After the service, the group moved to the Holcombe-Riverview Cemetery for the burial, then to the American Legion for a reception whose menu of sandwiches and salads recalled the many family and community picnics held over the decades on the island.

After the funeral, the crew, including family members, changed from Sunday best into their crew T-shirts (usually worn only for the Shad Fest haul seine demonstrations) and hip boots for the festival's demonstration haul. One crew member, Jack Marriott, a boyhood friend of Steve Meserve's, also brought black grosgrain ribbon to the crew's cabin, and crew members cut armbands, even for the four-year-old on the crew, to mark the demonstration haul as a memorial haul. Stepping onto the island that day, visitors saw the customary preparation for the Shad Fest: everything was a little spruced up, and caution tape on temporary fence posts marked out where the haul would occur on the side of the island, providing the crew with room to work and visitors with a good view. This weekend, however, one also saw a spray of funerary flowers, which a family member set atop the steps to the crew's cabin, marking the space as family space and signaling to visitors that this wasn't just any Shad Festival. At the start of the haul, Steve had me talk to the audience to tell them about Fred's passing and our memorial arm bands, in addition to giving the usual talk on what to expect of the demonstration. Knowing that in some years the crowds have behaved rudely, even booing if the nets yielded no shad, the crew felt the need to prep the audience for the special occasion, avoiding any potential unpleasantness that the family would have less emotional strength to weather. Fortunately, the orientation worked, and the crowd became a sympathetic part of the memorial.

During the memorial haul, the family's and crew's actions and comments marked the occasion as spiritual in a way that struck a balance between public and private rites of passage, between historical significance and personal grief. Some of the flowers from the grave were brought onto the boat, and when the boat was in the river, barely visible to the festivalgoers, Steve Meserve, now sole captain of the Lewis Fishery, scattered the blossoms from the bow of the boat. Some of these petals were caught in the net about a half hour later, drawing excited remarks and tears from the crew members. Comments also connected the weather to Fred's spiritual presence. Saturday's fine weather was interpreted as a sign of Fred's pleasure, but Sunday's stiff headwind and precipitation showed that Fred was testing the crew's mettle, or perhaps playing a trick. After the haul, crew intermingled with festivalgoers, this time accepting condolences as well as educating the audience about the Big Stories of tradition and environmentalism, as they would any other year.

What was particularly poignant about Fred Lewis's passing and people's narration of it, though, was not just the import of Time and Tradition, with capital Ts, but older, precalendrical time rhythms: the rhythms of spring. There was pronounced irony in the fact that Fred died just when "his" activity was at its most lively point in the year and just as the world was coming alive for the year. While not intending to exaggerate Fred Lewis's figure as a *Christ* figure, I do want to posit here that the juxtaposition of death and life the week he died was fitting because he was such a devoted *Christian* for whom the Easter story, with its juxtaposition of death and life, was paramount (see above for Pam Baker's discussion of her Good Friday haul with her grandfather). Thus, the flowers that were scattered during the memorial haul linked his life's work with springtime itself, a symbol that gained extra import, given the season.

Fred Lewis's funeral coinciding with the Shad Festival brought together the life of a man with the annual high point in the life of his community, which he influenced in several spheres, including the environment, education, and recreation. The import was palpable as residents took part in the rhythm of viewing, then funeral service, graveside service, and finally reception, which is traditional in Lambertville, and the regular fishing hauls that week served as an extension of the visiting hours at the Van Horn–McDonough Funeral Home. Indeed, some folks saw each other in both venues on a single night, since the family felt that Fred would have liked the fishing to go on as usual. That week, the island and the traditional funerary activities hosted many of the same conversations, resembling the process Cashman describes at Northern Irish wakes: "[T]he deceased is resurrected through the exchange of anecdotes that amounts to a character study. . . . Given form in narrative, the memory of the individual as embodied in anecdotes is as much community property as it is familial property" (2008, 92). Whether the public discourse took place in the sphere of the face-to-face community or in the newspapers, it was obvious that people were witnessing the turning of an era, a passage with symbolic impact: on the one hand, the old era was done; on the other hand, the fishery and its community had clearly survived the rupture with traditions intact.

This turn of the generations in the captain role—two generations at once—could have produced much more anxious uncertainty, but those aware of the day-to-day operations could have confidence in the new generation because of Steve Meserve's demonstrated commitment. Two anecdotes bolster the image of his lifetime commitment to fishery traditions. When Steve went to college and graduate school in Virginia, he missed the fishing season each spring, but during that time, his father's death made it clear that either Steve or his aunt, Sue Lewis Garczynski, would likely be next in line to lead the family tradition. After graduating, he knew that he needed to return to Lambertville, the Delaware, and the Lewis Fishery, which he would be "forever involved" in (Steve and Sue Meserve, April 22, 1996), a phrase he has used more than once when telling of the turning point. The realization defined Steve's role and identity for him and figures in an anecdote told on occasion by his wife. When Sue and Steve began getting serious as a couple, one day in the parking lot on the mainland after fishing, Steve very deliberately turned to Sue and said, "You know I fish in the spring." Sue took the full import of the words to mean that if she was signing on to marriage with Steve Meserve, she was signing on to having a husband who fished, with all that entails: not going away in the spring, late or lonely dinners, and a hectic home schedule six days a week, plus the responsibility of caring for the fishery and the island. Sue accepted the situation with full knowledge, and the anecdote signals the importance of commitment in yet another generation of the Lewis Fishery captaincy.

Continuance does not necessarily mean without change, however, as shown by stories about Steve Meserve that have begun to emerge as he develops as a captain figure in the public view. Steve's adoption of the role itself is unique because of the way the role was delivered to him. Great-grandfather Bill Lewis Sr. chose to step into *a* captain role, but he created the role of captain of *the* last commercial haul seine fishery in the region. Grandfather Fred Lewis had a role thrust upon him abruptly following another's death—twice, first when his father died, and then again when his brother died, leaving him as the sole captain. Steve also had a premade position and status to step into, but the stepping-in process was not so clear. For starters, the typical father-son, patrilineal inheritance pattern was disrupted by gender (his parent in the bloodline is female) and skipping a generation. Unlike Fred Lewis, who shared the role equally with Bill Jr. then two years later unexpectedly received the entire role, Steve played the junior partner in the captaincy with his grandfather for almost a decade before inheriting it entirely. On the plus side, this elongated liminal stage allowed him to learn more about leading a crew, caring for the island, and reading the river and the season before taking the reins entirely. On the challenging side, Steve Meserve had to defer to his grandfather in practical and honorary matters while piecemeal taking on the direction of a crew he had been a child in and graciously helping his grandfather relinquish a well-loved and central part of his identity.

Steve Meserve's successful navigation of his first years as operator required a special combination of personality traits, some of which emerge in stories about him. To be sure, Steve stories are just beginning to come forward, and he is a developing character in microlegends in which plot is predominant (see chapter 4). Yet, one story marks the start of Steve's reign as sole captain. In 2005, the first March after Fred Lewis died, Steve phoned the crew to start, as he had for several years before. Early into that season, one of those unusual but not rare April snowstorms dumped several inches on the region. Steve called up the crew to call them off from the night's fishing, announcing with delight in his voice that he created a new rule at the fishery: "If we have to shovel the path, we don't fish." In this story, the new rule is Steve Meserve's rite of passage: while before he led the work of the crew, he now had full authority to set rules without consulting anyone. This story appears almost iconic when one notices that while wielding authority, Captain Steve also, in a sense, takes it easy on the crew, a pattern that matches both changing times for the Lewis Fishery and the world around it, and Steve's unique personality and life experience.

As senior programmer for a large insurance company, Steve Meserve had little problem stepping into his grandfather's and great-grandfather's role in managing data for outside agencies. While continuing with the notebooks that suit the slime-flying nature of data collection at the Lewis Fishery, Cap'n Steve easily expanded his data management techniques to use spreadsheets, e-mail, and websites, both tracking and adding to the data spread among fisheries, scientists, and government agencies through the internet. This authority in a specialized area of the public sphere complemented the practical authority he had standing clearly at the apex of the crew's hierarchy. Lastly, his authority as the fishery's figurehead developed quickly when his mother and grandmother (the next two local owners of the fishery) and reporters conferred this status on him by making sure his voice led stories related to a series of significant floods from 2003 to 2006 as well as the perennial Shad Fest news coverage.

Although clearly wielding authority like his grandfather and great-grandfather before him, Steve Meserve does not have the same reputation for unbending high-mindedness and stubbornness attributed to several of the Lewis men. Undeniably stable and determined—to preserve the tradition, island, and shad—Steve Meserve leads with fewer commands and more consensus than his forebears did. Steve's word is final with the crew, but they also feel greater license to suggest another way of doing things, volunteer their preferred roles, or even whine a little about fishing in unpleasant weather. This last factor has developed into a joking practice between captain and crew—what can we get Steve to agree to? The crew members will even affectionately tease the captain, which would not have been attempted with any of the previous captains. Because of Steve's amiable and patient nature, microlegends of him getting angry at boaters heading for the net are particularly hilarious fodder for razzing. This happens frequently enough, that when asked about stories told among the captain and crew during the net-mending stage of the haul, Sue Meserve replied, "Well, and then there's just the general hassle Steve story," adding "I think each year is a chapter," and her husband agreed (October 3, 2009). The humorous and practical back-and-forth between captain and crew match Steve's tolerant personality, which also enabled him to straddle the leader and follower roles when operating the fishery with his grandfather at the helm.

In the end, consideration for consensus of opinion, self-deprecation, and easy-going personality characterize Steve's leadership, which are then balanced with the crew's respect for his considerable knowledge of the place and the activity, his formal status as captain, and their great affection for him. Steve and Sue also carry their community stewardship role into the offseason, when they make sure that crew family news (such as births, illnesses, and deaths) spreads to others in the crew. This caring, stalwart, and committed but not stubbornly unbending captain figure not only reflects Steve's personality but also makes a better fit for today's fishery, which is a wholly voluntary enterprise. Steve Meserve can't promise a wage, so he must find other ways to keep the crew full and committed. He has had to abandon the commanding presence of past captains for a friendlier persona, adding to the enjoyment of the activity and the sense of purpose offered by the Big Stories.

Yet, even in change, we see the same themes arise in the Big Stories and the captain stories: commitment and faith; intelligence as a naturalist and a leader; good-hearted, upstanding character. Cashman argues that "the very project of focusing attention on individuals, through characterization, may ... provide us a glimpse into collective concerns" (2008, 220). These "collective concerns" of the community include the kernels of the Big Stories: the stewardship of the natural resources and of the community itself, including its history and the moral values of civility and commitment. In sharing the captain character anecdotes, tellers and audiences not only create a knowledge base of collective memory and local oral history but may also push each other to take on the concerns of the captains and use them as models to some degree.

Cashman calls attention to the special power of character anecdotes to express communal concerns, for "the anecdotes about [characters] bring narrators and listeners together in circles of participation that create community while simultaneously representing those who comprise and may symbolize it" (2008, 222). In captain character anecdotes, their leadership and other positive characteristics are emphasized, while negative details may be omitted altogether or given a positive spin depending on the time available for the telling, or the teller's comfort level with the audience. Telling a negative anecdote without adding a positive one to balance it may show that the storyteller recognizes the audience as a member of the community who knows the larger context of that captain's life. That is, when Muriel Meserve tells me the story of Bill Lewis Sr. not sending Fred Lewis to college, she does so knowing that I understand that Bill Lewis Sr. also competently, personally, and deeply cared for his three motherless children, and will relate the larger context to my audience. Similarly, Bill Lewis Sr.'s and Fred Lewis's "stubbornness" or "orneriness" could simply be presented by the terms "certainty" or "dedication," but instead family members often choose to take a playful attitude with many audiences—even with the media on the scale of the "imagined community" (Andersen 1991). This risky show of intimacy reflects humility as well as investment in future relationships.

Such an investment in community building is not limited to the content of information sharing but also involves the process of sharing information. This function compares well with Cashman's discussion of his subject matter: "[T]he local character anecdote facilitates the dialogue of the community as network and the community of the social imaginary—through the anecdote the two are mutually constitutive" (2008, 256). What people say about the captains as representatives of the community creates an image of the community, while the dialogue—the communication process of *telling and listening*—creates relationships and thus community itself. Passing on stories of captains has a place in this communal process, but as the next two chapters will show, when tellers and audiences turn personal experiences into legends in the context of everyday storying, day-to-day living and narrative stewardship are even more inextricable.

FOUR

"Were You There When . . . ?": Microlegends

We knew that something big was coming in the net, but when the net started to fold in half such that the lead line and cork line met like a cinched bag, the ferocious splashing was not that of a nice load of shad or even spikey, rancorous catfish. No slime and scales flying, just churning water . . . and claws . . . and teeth. We had caught a beaver, the one who had been felling saplings along the island's outside path. For weeks, we had shivered at seeing its path of destruction as we pulled the boat upriver at the start of each haul, kind of like the feel of the Jersey Devil stories we'd heard growing up. Now we encountered its huge size. All of us terrified, crew member Tim Genthner leapt into action first—thus earning the moniker "Beaver Boy." He stepped on the top of the cork line to enable the beaver to hop over the net, like misguided rowers sometimes do. However, up and over is not in a beaver's nature. The thrashing beast kept trying to dive through a foot of water into cement and cobblestone, until Tim finally went against a shad crew's instinct and sacrificed the haul, pulling up the lead line just moments after the nightly "keep the lead line down" chorus. The animal swam away, and "The Night We Caught a Beaver" became a microlegend by sundown.

What and When: Definitions and Storytelling Settings

One of the particular narrative types found on Lewis Island, "microlegend," is akin to legend in that it is told as true about real-life figures, but the scope and context are local, emerging from happenings at the fishery or close by, most often within living memory. The microlegend type may overlap with character anecdotes about captains, and both may support the themes of the Big Stories. In contrast, "everyday storying" (to be covered in chapter 5), emphasizes the most ordinary and quotidian aspects of telling personal experience narratives. While everyday storying focuses more on the mundane, microlegends lift experiences out of the mundane for special attention.

Citing Elliott Oring's work with narrative, Ray Cashman (2008, 109–10) draws our attention to the fact that legends are not formally much different from the surrounding conversation, because content is more important than artistry. Frequently the microlegends are "micro" in form, with artistry second to content, values, and truthfulness. In his classic article "The La Have Island General Store: Sociability and Verbal Art in a Nova Scotia Community," Richard Bauman argues that "although telling yarns [exaggerated wholly or partly fictional stories] was an important component of the process of establishing a social identity, being a good storyteller was not itself a significant identity feature" (1972, 337). Similarly, on Lewis Island, people connect themselves to the island, haul seine shad fishing, and Lambertville through storytelling as a way of establishing their own identities, but being an artful storyteller is not as important as having knowledge and experience of the community, or as important as enacting Lewis Island values. Indeed, a particularly artful storyteller can lose status if the storytelling appears to be too professional or practiced, dividing teller, topic, and listeners from each other.

Several scholars of culture have studied value systems that require connection between authenticity and tradition or the past (Bendix 1997; Cashman 2008; Hobsbawm and Ranger 1983). In such systems, the professional and competitive storytelling events that emerged with other folk revival activity in the late twentieth century may be considered a "new" thing, and therefore less authentic than an aesthetic that embeds storytelling in everyday conversation, keeping teller and listener within reach of each other as well as linking the past, the place, and the people. By telling a particular microlegend, one asserts status through connection with the past. That status can be greater or lesser depending on whether the teller witnessed the event, how long the teller has known of the story, who originally told the teller the microlegend, and what the first and subsequent tellers' relationship to the story was. The relative position between teller, time, and place (Lewis Island) can also enable the teller to share status with an audience member who hears the story for the first time. Narrative stewardship occurs primarily through sharing of stories that empower both tellers and audiences to conserve community resources.

Talking of flood stories emerging in Indiana, Barbara Johnstone labels the barest outline of plot a "narrative core," which is then "fleshed out" with characters, settings, "and elements that underline the story's relevance and point" (1990, 23). Their repeatability and important content are what make them legendary. Part of the "micro" quality of the fishery's microlegends is the fact that these discrete stories at the fishery, like the stories Johnstone presents and discusses, center on the concerns of a small community. In my interviews, informants assuming my group membership condensed the narrative core into phrases that approximate titles, such as "The Night We Caught the Beaver," "The Night the Brail Broke," and "The Longest Night of Fishing." These stories can be merely referred to when former crew members return to the island for a visit and wish to reconnect with crew members from their era. "Old crew" can also share these stories with new crew members, friends, and customers. Crew member Keziah describes a process in which adults in the crew and family collaborate in telling stories to children. For example, she says of captain Steve Meserve,

> He'll tell stories . . . to us, but usually he'll say something that would trigger a story that either you or Sue [Meserve] will tell me and Adelaide [Keziah's younger sister]. Like he usually isn't the one that actually tells the story, but he'll trigger something about it. He'll say something vaguely, and then we won't get it, so then you guys will tell the story to explain it. (October 3, 2009)

The stories Keziah refers to here might be parts of fishery history but may also be discrete microlegends. Both island children and visitors of all ages are targeted for telling microlegends, using the excitement and novelty of the story to get people interested in passing oral history and preserving community.

Wonder Stories

There are two basic types of microlegend cycles: surveillance stories (to be described in the next section) and wonder stories, which are narratives about unusual happenings. The story of "The Night We Caught the Beaver" exemplifies the wonder stories that catch the crew's attention, and although beaver have been caught since, the story of the first time this crew caught a beaver remains *the* beaver story. Although fishermen are known for exaggerating stories (particularly regarding the proverbial fish that got away), narrators at the Lewis Fishery strive to remain truthful in keeping with the values of civility and the integrity that is central to the captains' character anecdotes. Indeed, the fishery family's Christian faith in the Ten Commandments, including "Thou shalt not lie," allows for joking and "tricking" but does not extend to making the crew or captain look important by exaggerating the size of a catch. That same faith is anchored in the idea of the natural world reflecting a divine creation; it is the truth of the haul, not the extent of imagination, that inspires awe. Rejecting the fisherman's yarn-telling tradition also has practical implications for the crew members' stewardship role. That is, they could not effectively report environmental problems to government agencies and enlist their help in protecting the river and shad if their honesty were doubted. The fishery's alliance with the press in getting out the Big Story of the environment also requires strict adherence to the truth. As reporter Renée Kiriluk-Hill says, "The knowledge is a big part of [the captain's role], because otherwise it just comes across as fish stories" (June 3, 2011). Clearly, "fish stories" are not valued by the paper, and exaggerated wonder stories could compromise the ability of the press and the fishery to raise public awareness about environmental issues.

The quintessential wonder story is perhaps "The Longest Day Fishing," an event that happened before any of the current crew was born. Fred told it this way on March 1, 1996: "The longest day I ever put in was from midnight Sunday night until four o'clock Wednesday morning. We fished a haul every hour." Fred Lewis enjoyed framing the "longest day" as lasting multiple days, as was characteristic of his sense of humor and skillful turn of phrase, and this story appears in Bruce Stutz's *Natural Lives, Modern Times: People and Places of the Delaware River* (1992, 228–29) and John McPhee's *The Founding Fish* (2002, 256). When asked to recall fishery stories that stand out to her, Keziah replies, "Well, there's the one that people mention a lot when they, I can't remember it exactly, but they were fishing for a couple of days straight, or something, 'cause the shad were so good. And so, they would sleep, and they'd take shifts and just keep going" (October 3, 2009). Condensing the story in telling it to someone she expects will know the story, she picks out the important parts: that the fish were plentiful, the fishing kept going from one day to another, and the crew cooperated to perform this feat of endurance. The triumph here is a combination of good fortune, strong backs, and good teamwork under conditions when a sleep-deprived crew could become testy.

Other wonder stories emphasize just the fortune—good, bad, or just *strange* fortune—of what appears in the net, such as in "The Night We Caught the Beaver." In his recollection of key microlegends, crew member Dan Tuft recalls the same stories that other crew members of his era do: "The night they caught the beaver, and I wasn't there. . . . Whenever there's sturgeon . . . , which I think has been maybe . . . twice . . . since I've been there? Maybe just once? . . . The time that we caught that huge grass carp that looked like a monster" (October 3, 2009). Different narrators hearken back to strange hauls they witnessed themselves, and for Muriel Meserve, that recalls the night the crew caught a muskie:

> Just because it's so big and so, so ugly and so rare. And *everybody* was there with their cameras taking pictures of this, you know. And they wanted to get it back in

the water, because you can't keep them, but also wanted to make sure that everybody got their pictures, so you were kind of torn between the two. So let's get the pictures, but don't let the fish die here either, you know? [laughs] (October 5, 2009)

Here, Muriel catches the practical tension in the stewardship role. First, the crew cannot control which fish appear in the net; they can just make sure they safely release those they cannot or will not keep. They need photos to help tell the story, attest to its truth, record the local happening for local history, and report the details to authorities who monitor the environment, but they also need to save *that* specific fish and get it quickly back into the water. In recent years images have proliferated on the internet, and on one occasion this tension was heightened when a vocal critic erroneously assumed that the time it took to snap a photo of a fish endangered the fish. Thus, not only must the crew actually protect the fish, but they also need to carefully manage the impression of protecting the fish. This brought amusing results the night the crew caught an almost four-foot-long grass carp in 2007, the event Dan mentioned in his interview. Pam Meserve Baker identifies the grass carp story as one of the children's favorites (October 4, 2009), which perhaps is so because it was a child who resolved the tension between protecting the fish and recording the natural oddity. While rushing to take photos before the fish was taxed, the crew reached for the best readable measuring device close at hand: Adelaide, a six-year-old who happened to be the same length as the grass carp. So important in crew memory is Adelaide's role as yardstick that the story is often called "The Night We Caught a Fish as Big as Adelaide."

The occasional sightings of sturgeon show the tension sometimes inherent in the complex relationship between narrative and stewardship. On January 31, 2012, the National Marine Fisheries Service listed the Atlantic sturgeon as an endangered species under the federal Endangered Species Act (*Today's Sunbeam* 2012), but before their numbers dropped to dangerous lows, sturgeon were included in the list of food fish the fishery is allowed to catch, keep, and sell. In fact, they were a particularly plentiful population at one time. However, when sturgeon numbers dropped, the Lewis Fishery refused to keep them even though it was legal to catch them, and if one ever appeared in the net, freeing it was the highest priority, done with great speed and little-to-no human contact. A portion of the catch was released if doing so would give the sturgeon a less stressful release back into the river. If someone needed to help the sturgeon into the river, the captain would be the only one to touch it, handling the fish as gently as he would a newborn kitten. The sturgeon inspire awe because of their unusual, ancient appearance, low numbers, and sometimes huge size—they can grow up to fourteen feet and eight hundred pounds (*Today's Sunbeam* 2012). They also inspire joy, because their presence shows that the river is very clean as a result of environmental activities owing in part to Bill Lewis Sr.'s great efforts to catch the authorities' attention. Thus, their presence in the river is a confirmation of the Big Story of environmental stewardship.

In 2013, three smallish sturgeon appeared in the net at once, instantly becoming a wonder story among the crew. On earlier occasions, sturgeon might be spotted, but it wasn't until multiple sturgeon appeared together that it was clear that they existed in at least small numbers on that stretch of the river. Nevertheless, this microlegend is rarely discussed outside the crew, and when the occurrence is referred to in a semipublic setting, the narration will include a recounting of how quickly and gently the sturgeon were released, include the preferred wording that sturgeon were "spotted," and/or will include a caution about not letting the story get out widely. There is essentially no physical risk to sturgeon caught in the net, and their special treatment—both physically and narratively—seems to have almost ritual import, beyond what is practically needed. At the same time, the stories are limited to protect the fish from thrill seekers who might hear the stories and seek out the rare sturgeon, posing a real threat.

While effort is taken *not* to widely retell wonder stories about sturgeon in the net, there are a couple of very tellable stories about extremely large sturgeon that have been observed outside the net. In an interview, Diane and Ted Kroemmelbein share two stories of oversize sturgeon from two generations of shad crew. When asked for repeated crew stories, Diane thought of one her father, Johnny Isler, repeated to the family and other crew members:

> I don't remember when it was. . . . But they used to talk about this one time they were just, you know, they were hauling the net or whatever, and they saw these two things coming up the river, and they looked like torpedoes, and they were humongous sturgeon. They said they must have been six feet long, and . . . it just caught all their attention. [To her husband, Ted] Do you remember my dad [Ted: yeah, yeah] talking about that? I wasn't there, but . . . I don't know if that's something Steve would remember or not. (June 15, 2013)

Subsequently, using references to her daughter Kelly's youth and Steve's memory, Diane places the story possibly as late as the 1980s, a time of relatively good fishing, but when Steve was the young guy on the old crew. Ted brings the conversation closer to the present by adding his own experience as landsman in the next crew, telling this story more slowly and dramatically than other narratives he shared:

Now, one time, and I never saw what it was, but I was fishing in higher water, little bit higher water than what we have been, but you know, slow water, but just a little higher than what it is now, and it was near dark. It wasn't *dark* dark. It was lighter than this [the lighting at the time of interview], which is dusk, twilight. And I saw a wave coming up close to the shore, and it hit the net, and it pushed the net backwards for a couple feet, then it went underneath it and took off and le—, and up the river sedately, leaving a wake with it. I don't know what that was! My guess is that was a big sturgeon. I didn't see it, you know, because they're brown and blend right in with the bottom. You wouldn't see it, unless it flipped over and had the light brown bottom. You wouldn't know what it was, but that's my guess. (June 15, 2013)

As is common with legends, Ted uses a reference to the current river height and current lighting to connect the story with the present and his audience, for he accurately assumes that people accustomed to spending time around the river will be aware of its height at any given time. Through his telling, Diane engages and bears up the story, agreeing with the difficulty of positively identifying sturgeon under water, the likelihood that Ted had spotted one, and the emotional impact—"thrill," she calls it—of being in a sturgeon's presence. As Ted summarizes the story as a closing, he emphasizes the fish's power: "That's the only, you know, explanation I got for that. Pulled the net back! Stopped. Stopped the net, you know; that's powerful. Whatever it was was very powerful. About a second, and then it went under the net" (June 15, 2013). Saying that the creature stopped the net not only gives factual (sensory) evidence of its presence, but this core feature of this wonder story also attests to the power of the being, for the powerful river constantly pushes the net in the opposite direction. Ted's slow pace and serious tone throughout the story, emphasis on mystery of what he believes but cannot know for sure, and testimony of something powerful and invisible all strike a note of awe, almost of the numinous, which often attends wonder stories about the river's power and unpredictability.

Another awe-inspiring story, one of the most joyful wonder stories told by older generations, is "The Night the Shad Came Back," which occurred the season before Bill Lewis Sr. died. This microlegend sometimes serves as a coda to that captain's instrumental role in the Big Story of the environment, as in this telling by Fred:

[In] 1961, Dad died and he was always certain that the shad would come back. That there would be shad back in the river. . . . He did live to see it. We caught us a haul of twenty-two one night, I think it was, about 1960. And that was a big haul then, so we took it up to him, his house on Clinton Street [on the mainland, a few blocks away], and showed him the fish. So the next night he was [laughs] down there. (March 1, 1996)

The year is likely to have been 1961, since the total catch for 1960 was six shad from nineteen hauls on twelve different days (an average of 0.3 shad per haul), but the next year the total catch jumped up to ninety (an average of 3.5 shad per haul). This story not only points to Bill Lewis Sr.'s important role and the importance of being *at* the fishery, but also to the wonder of a considerable haul after a dearth.

Muriel Meserve and Nell Lewis recall a different wonder story occurring earlier in the twentieth century's dark years, and in this story, they refer to the experience of hearing the stories secondhand, then move quickly to the public excitement and increase in visitors after the later happening described by Fred. Prompted to recall stories told repeatedly along the lines of "the night X happened," Muriel first tells this story, which likely happened in the 1940s: "Well, back when they were having trouble catching the shad, . . . [we] had a party for Memorial Day, and . . . they ran out in the river, and Poppy said . . . , 'Let's make a haul! Let's try and see if we do have any.' And they did . . . make a haul, and they brought in, oh, an awful lot of fish, of shad" (October 5, 2009). Nell Lewis adds to the story: "They continued in fishing until June the *tenth*, because that was the date, uh, that was the last day." Here, Nell points out how unusual it is for the season to last until the last day allowed by the license. Usually the season ends some day in mid-May, when the daily catches suggest that the bulk of the run has already passed Lambertville. Muriel punctuates the wonder in the story: "They just hadn't been catching *any*thing, and then fftt! [Charlie Groth: there it was!] There they were!" When asked about the reaction on the bank to the unexpected shad, Nell Lewis shows that she was not a witness but rather among the first in the audience for the story: "Well, I remember being told, but I don't know; I wasn't there yet at that time," and Muriel agrees, "No, I don't—, we heard about that one *after* the fact," and laughs. She continues, "I was *young* then, really young. And so we must have been living there, but we weren't, obviously not down watch—, probably were cooking [laughs], not down watching, you know, because you did get tired of watching these hauls with nothing in them." Directly after telling this story, she then immediately presents the contrasting excitement of the years after "The Night the Shad Came Back," the event of Fred's wonder story: "Although in the sixties *everybody* watched, and, you know, those early sixties, and . . . for a long time they would say, the night when we caught the four hundred, you know, the night when we caught the six hundred, and you couldn't pull the nets in. You had to just *stand* there with them, and [the nets] started to tear, and the whole thing that went with *that*" (October 5, 2009). Here, she moves directly from the good times of the 1960s to later occasions between the mid-1970s through the early 1990s, with healthy annual catches ranging from 1,120 to 4,790. When asked how many shad would be in a "good catch," Fred Lewis, speaking from the perspective of over eighty seasons, replies,

> Oh, a good catch would be a hundred. We've caught as many as, well, we've *counted* as many as four hundred before. That's the biggest haul we've ever counted. But we've caught and hauled, in the last, well, in 1992, we had hauled and we had five or six, seven hundred fish. We had to let them out; we couldn't bring them in.... One night we had a big haul and two fellows were pulling the cork line, the line of floats. They had to stand on it, stand there and just push the fish out this way, while we were taking 251 out of the net. We took 251 out, put [them] in the tank where they were tagged. And then the rest of them we just let go. We felt sure we had over seven hundred in that haul.

Sue Meserve frequently recalls this particular night as one of her favorite wonder stories, as do Ted Kroemmelbein and Steve Meserve. Nothing close to such a haul happened for twenty years thereafter, but the 2013 season became a wondrous season with hauls in the twenties again and a season's catch over a thousand, after two seasons of 52 (2011) and 196 (2012). Although in the 2014 season more than 400 shad were caught, that was still under half the catch of 2013. After three more lackluster years, 2017's catch totaled about 1,300 shad, likely because the 2013 fry were spawning for the first time. While the general public tends to see every year as a chapter in a story about the demise or recovery of the shad populations, people who know shad realize that the population is cyclical. Therefore, where others saw the 2011 and 2012 seasons as grave disappointments, the family and crew expected them, focusing instead on the fact that the 2017 catch confirmed the positive surprise of the 2013 catch. The excitement of early hauls in 2013 and 2017 recalls Fred's microlegend of the shad coming back in 1961 in a haul of twenty-two.

While more humorous than joyous, a last cycle of wondrous catch stories overlaps with wonder stories about overcoming challenges in the fishing but instead includes stories about hauling in human-made objects, and these stories often emphasize feats of strength. They are frequently condensed to their titles and listed in a chorus by crew members for curious listeners: "The Night We Caught a Telephone Pole"; "The Night We Caught a Lawn Chair"; "The Night We Caught a Sign." Hauling out a telephone pole obviously requires a full crew pulling against a river as well as an obstacle that will likely rip the net and lose the fish. Other obstacles, such as the lawn chair or construction sign from a previous bridge repair project, may not be heavy per se but nevertheless will try the patience and strength of the crew, for these objects act as anchors, catching each rock on the river bottom—and again not catching fish. Unlike the beaver story, these trash-hauling stories recall the floods that dislodged the objects and brought them downriver. Thus, these stories remind people that however strong the crew is, the seemingly casual river is even stronger. The narratives also point to the crew's role in river stewardship, hauling out the trash that could harm the species in the river or hurt the people enjoying the river. Here, the dramatic tension between the crew and river resembles the ironic relationship between the crew and shad: the crew struggles against these natural forces while simultaneously taking on a steward's responsibility for both river and shad.

Overcoming mishaps—such as failing equipment—in the course of fishing forms another cycle of microlegends, and these are also recalled with phrases that begin "The night that ——." Oars breaking are commonplace enough that they are not bestowed with microlegend status unless paired with some other situation, such as happening upriver while a Shad Fest crowd of a couple of hundred stand waiting downriver for the hauling demonstration. Here, teamwork and ingenuity are highlighted—Steve sending Sue to make a new oar during the demonstration by switching out the shaft while sending me down to explain the situation to the crowd and entertain them with information until the new oar was made and the haul could continue.

In the late 1970s when Steve was still in high school and fishing on his grandfather's crew, he recalls one time when the crew and captain made errors in judgment, making a big mess of the net (March 7, 2015), which was an extreme rarity once Fred came into his own as captain. Steve sets up the story by explaining that the "water was up," and, whereas they should have fished with only 50 or 100 yards of net, they used about 150 yards. Rather than stopping to remove net, anticipating that the water would go down the next day and they'd have to put it back on again, Fred Lewis decided that they would "just fish this and fish it like a shorter length of net" (Steve Meserve, March 7, 2015). This particular length of net was attached to aging hemp ropes rather than the nylon used now, and with the intense pressure of the high, fast water, the lead line broke in several places as they pulled it in. Fred yelled to Johnny Isler, the landsman, to tie up the other end of the net while they tried to bring in the rest of the net safely. Steve recalls that the crew was safe, but "we destroyed the net." In telling this story, Steve repeats that the disaster happened in part because there was more pressure to work efficiently and fish more hauls thirty years ago, because they were still able to make money back then. Today, Steve would more likely have the crew stop to take off net or not fish at all that night, but, he says, when fishing was still profitable there was "a greater range of potential problems than we put ourselves in today" (March 7, 2015). Thus, the story is wondrous because of the danger and destruction and also because it became unusual: not only because of Fred Lewis's misjudgment but because that experience and changing economics have since caused captains to assume less risk.

Stories of failing equipment from the twenty-first century then become all the more exciting because they happen in spite of more conservative choices. Two especially exciting microlegends are "The Night the Brail Broke" and "The Night the Rope Broke," which at times become confused with one another. In the first story, the aging pole (brail) that holds the net open snapped in two, while in the second story, the brail became detached from the rope. In both cases, not having a full crew created a dangerous situation for the net and for those crew members on hand. When Pam Baker was asked for stories that she was part of, rather than ones she'd heard from earlier generations, the first one that comes up is "The Night Tim Saved Us," which she references and narrates in a dramatic voice:

> When Tim came and saved us because we broke the, you know, and when Steve had to go get the pole, and it's those kinds of stories because we're part of it. But the thing is the kids were part of it too, so it's great. So they have, even if they were very young, it's like, no, don't you remember when . . . ? You know, . . . when we got the three sturgeon, and . . . it's very factual kind of things. And, uh, but, yeah, yeah, the Tim one is the one. "Oh, Tim *saved* us that night! He came in on the boat and got us unhooked and . . ." You know, that was a great night. (October 4, 2009)

Here, Pam links her experience tightly with her role as the mother of children in the fishery's line of inheritance. She clearly uses the stories to build their interest and knowledge, and of course she willingly retells their favorites. When recalling stories, Keziah also distinguishes between the stories before her time and the stories during her years on the crew, carefully noting the events she witnessed directly. When asked for stories during her time, she responds, "Well, I wasn't actually there for that night, but about the beaver that we caught, and there was the one where, uh, the net broke, and the net went past the bridge, and I was there that night. That one we talk about sometimes" (October 3, 2009). Here, Keziah refers to pinnacles of excitement that are consistent with others' estimation, and "The Night the Brail Broke" was mentioned prominently by each of the current crew when asked for wonder stories involving trouble, which Dan Tuft calls "legendary mishaps" (October 3, 2009) and Ted Kroemmelbein calls "near calamities" (June 15, 2013).

Kroemmelbein explains how he has learned to handle the net over time through experiencing the river, and how his choices throughout the haul can ease the crew's work during the final bagging-up process. Yet, as he tells of his success, he quickly shifts to the opposite experience, when things go horribly wrong and the landsman is left quite vulnerable. When successful,

> it looks like the river's doing it, it looks like none of us, you know, they're just out there rowing around and come back in, and the guy drags the net down and, you don't do that right, and it's all screwed up, you know, and things break, and, you know, we've had a lot of near calamities. Ohh, when the net breaks in high water, and you can't hold it, you know, no human can hold it. You need about fifteen guys sometimes to hold that thing. (June 15, 2013)

Ted then launches into the story that crew members commonly refer to as "The Night the Brail Broke":

> I remember one time from the top of the island, when we got out, uh, right off the bat, with a lot of net [laughs], it broke on the boat side, so I had all that net out there with high water running about twenty miles an hour or fifteen miles an hour. And we only had about five guys total, three in the, two in the boat, and then one down there waiting. It wasn't a full crew either. And I couldn't hold it. I couldn't.... And we start off, you don't have much slack left.... Well, as soon as I was pulling on it at that point, with not much net, it broke right there, and I had to run down with it all the way. And they couldn't get the boat in. And, and I couldn't get enough to tie it around a tree. I couldn't pull it! I couldn't move it! It was too strong.

After a pause in which I try to ascertain whether this is the night the brail broke or the night the crotch rope on the brail came loose, Ted resumes the story from his perspective:

> And, thankfully, when I got down by the house, there were two guys leaving the island walking back across the bridge and I started screaming at them, [asking] if they could help. And I'm coming down at a really, really fast walk. Gone to the end, I would have had to let go. And they came back and between the three of us, we got enough and put it around a tree. And the end of that net was on the pier [of the Lambertville–New Hope Bridge]. And we had to go out and get it. We had to go out and get it into the boat the next day.

I try again to ascertain whether this is the story that Pam Baker calls "The Night When Tim Saved Us," or whether Pam's story is about the time the rope came loose. When I ask Ted about getting the net, he replies, "Timmy. Tim came to help. He got his boat and, and we went down there and got it with his, jet boat, and pulled it back up with the jet boat." Ted is unable to recall whether Tim's involvement happened that night or the next day, and quickly moves attention back to the drama of the story: "But it would have been gone. You know, it would have been down in Trenton someplace, and that would have been a mess." With this reference, Ted makes the story geographically bigger, for Trenton marks a natural border for the river, which widens and deepens significantly there and becomes tidal. In fishery stories, Hancock, New York, the farthest point north where shad spawn, forms the northern boundary, while Trenton forms a southern boundary, with occasional references to the Delaware Bay and ocean beyond Trenton. Mentions of Trenton function not only as references to the state capital and government authority (as in stories of Bill Lewis Sr.'s testimony), but also to something of a symbolic distant boundary, the edge of the fishery's control.

Ted then recounts the story a third time, again emphasizing the effect on the crew and himself, in particular:

> [When it started] I was still at the top of the island. So they had gone out and we started putting tension on it, and holding land was just straightening [the net] out, and then it broke. And I only had about five foot of rope behind me and that was it. Off to the races, all the way down.

My unsuccessful attempts to untangle the two stories of broken nets suggest that to the teller, the technical story is not as important as the story of the crew's peril and how it saved the day. Looking back at Pam's retelling, one notices that she breaks off her sentence and leaves out the explanation of what actually happened to the equipment to cause the problem: "When Tim came and saved us because we broke the, you know." Here, she assumes that her audience (me) knows what happened, and her omission is consistent with the way Ted's and Dan's accounts downplay the practical details. The main point is not that the equipment broke, which is to be expected in an operation with antique and cobbled-together equipment that is reworked or mended almost every night. Rather, the most important comment about the equipment is that it is one of a kind and cannot be lost entirely. The bigger story is that crew members (who are just regular folks) and even bystanders (who expected to relax) all worked together to save the net. As Ted tells of being alone in his pressure, he also tells of not being alone: the rest of the crew and strangers were there to support him. In Pam's telling, the damage to the equipment is entirely eclipsed by Tim's involvement. Both crew member and bystander, for he was on his day off, Tim saw that the crew was in trouble and brought his efforts and resources (a motorboat) to rescue the historic operation. These are wonder stories about peril, yes, but also moral stories about humanity on the island that bolster the Big Story of civility and community.

Indeed, some of the microlegends are specifically about people *not* on the crew, but rather about outsiders who endanger the fishery (and who will be discussed below in relation to surveillance stories) or about customers who support the fishery. Generally, discussion of customers relates either to the cultural diversity on the island or sometimes to an extreme desire for shad. In my first interview with Steve and Sue Meserve (April 22, 1996), they told me of a customer who so loved the fish that he had the Lewis Fishery freeze shad and fly it to Texas, a story Fred enjoyed telling, too. Pam Baker also refers to this story and connects it with her family's can-do attitude in serving their customers. Interviewed again in 2009, Sue Meserve recalls a similar oddity involving customers: "The guy came and wanted herring and we said, 'You gotta take 'em all,' and it turned out to be a trunk-load full, and we knew that because we filled the trunk of the Cadillac" (October 3, 2009). Here, the stories have the spirit of wonder stories about unusual catches, for the nexus of wonder is something out of place: a shad in an airplane, a shad in Texas, a herring in a Cadillac. This latter story is all the more amusing because while the bad odor of the fish is very much part of the fishery, fine cars are not. Despite their own dedication to the fish and fishing, crew members are not so dedicated that they would sully a Caddie. Such behavior from customers is the stuff of wonder stories.

Surveillance Stories

What I call "surveillance stories" make up the other half of the microlegend genre on Lewis Island. In these stories, stewardship of the island property itself is central, and the Big Story of stewardship contracts from mythical proportions to legendary proportions. These narratives report unusual happenings discovered while family members, aided by crew, neighbors, and other island visitors, collaboratively keep constant watch for threats to the island and island structures. These microlegends typically fall into two categories: stories of badly behaving people and stories of a badly behaving river. Both threaten the island and fishery, but while the flood stories (discussed below) are broadcast far with the help of the media, the people-acting-stupidly stories are mainly told within the fishery community. The stories about bad behavior reinforce values of civility but are told more privately so as not to uncivilly and unduly humiliate others. Knowing that I am to some degree "telling tales out of school," I share Ray Cashman's discomfort (2011) when he relates some of the more embarrassing character anecdotes and disguises identities. Fortunately for me, these stories of vandals and others who behave badly separate names and identities from perpetrators as part of the narrative norm, making it easier for me to share vandalism stories while I highlight the tension between the overarching values of decency and the protection of the island and its traditions.

Because of its accessibility and remote feel, Lewis Island has experienced its share of vandals over the years. Their behavior is carefully watched with some degree of resignation that opening the island to well-meaning and well-behaving visitors will also provide access to disrespectful people; a few pot-smoking or bottle-swigging youngsters have found refuge in the woods, just as did the invited young missionaries. On a daily or weekly basis, the fishery family, their neighbors, and people who regularly walk their dogs on the island keep an eye on the property. They report to Muriel Meserve their observations of littering, campfire ashes, theft, and property damage. While these mild episodes fall under the heading of everyday storying (see chapter 5), dramatic confrontations with people behaving badly rise to the level of microlegends.

Pam Baker tells a story about her grandfather's handling of vandalism in the early 1970s that highlights an intersection of righteous anger and teaching decency and responsibility. Echoing the concern for the difficult-to-replace nets endangered in the wonder stories of calamity, this story starts with the loss of the fishery's boat:

> Kids took the old boat that they had and took it down the river, and . . . I think it was the three-seater, the last wooden boat. . . . My grandfather did not . . . as a rule . . . yell. And he was very upset. He made them pay a hundred dollars, four boys, so they all had to pay, and at the time it was [a lot of money], and I just remember how he dealt with it. . . . The parents, I guess, wanted to pay for it. He was like, no, the parents aren't paying for it. The kids have to pay for it, 'cause I think the dock went too. (October 4, 2009)

Pam first emphasizes the value of the boat. Not only is the boat at the center of an operation without much cushion for loss, but it is also a very special boat: the last actual *shad fishing* boat, which enabled the crew to get out into the river with more rowing power, and which was followed by two shorter, harder-to-maneuver boats designed for other purposes and adapted. The end point is Fred's insistence that the children take responsibility for their own actions. Not only would this shrewd strategy more likely spread the word through the community that abusing the fishery's property had real and direct consequences for underage vandals, it also fit the Lewis family's commitments to civility and educating children.

In the center of the story, Pam calls attention to her grandfather's expression of anger. She does not make it completely clear whether he yelled on this occasion—whether "very upset" is a euphemism for yelling—but it is clear that if he went so far as to yell, it would have been out of character, confirming that this experience genuinely tried his patience. In fact, at other points in this interview, Pam repeatedly describes her grandfather as remarkably "calm," extending even to the manner of his death. She attributes her grandparents' calmness to their faith and believes that the "calm factor" suffuses the island's atmosphere and drives their success with children and youth on the island, whether they be Boy Scouts, visiting missionaries, their own foster children, or their grandchildren's party guests. She recalls, "I had Halloween parties over there that curl my hair now thinking about it, but my grandparents never freaked out about these things where I'm not sure I wouldn't freak out. Always very calm" (October 4, 2009). A high school guidance department employee herself, Pam connects the special atmosphere on her grandfather's island with the discussions of "civility" going on in educational circles at the time of the interview, and thus with the Big Story of civil behavior.

As crew members of all ages tell surveillance stories of people behaving badly, the captain shouting, yelling, or hollering (or *almost* speaking thus) become repeated elements that separate the incidents from other happenings on the island. Most of the stories involve people in pleasure boats headed toward or actually tangling in the net, which will imperil the particular haul if not cause extreme damage to the net. Such events have a cultural context as well as a legal context involving fishing rights. First, for two centuries the fishing nets have had the right of way on the river. Today, the FAQs section of the local Swan Creek Rowing Club's website (2015) instructs that rowers "may not row during darkness or during shadnetting," making reference to a "fish flag" that is posted to remind rowers to look for the nets around Lewis Island.

The caution to the rowers does not reflect the greatest danger, however, for a short scull could "hop" over the net as the geese do, and even as the empty fishing boat does right before bagging up during every single haul. The real danger is from motorized boats and personal watercraft (called "Jet Skis" after the brand but meant generically). Ted Kroemmelbein refers to a couple of specific threatening situations, then describes close calls as a type, retelling a general story that holds much in common with specific microlegends:

We've had some really, really, really, really close calls where I'm holding the net, and I'm ready just to let the thing go, 'cause this guy's gonna hit the net and the prop's gonna go. And, it's gonna yank me right off into the water and tear everything up, you know. And, and they stop . . . and he's out there [yelling in Steve's voice], You gotta stop! You gotta stop! Stop! And waving their arms and, you know, they saw it at the last minute and shut down, and . . . the net's here and the thing's like [makes the sound of a motor slowing down], just coasting up to it. Ooohh. I was ready to let go. (June 15, 2013)

After establishing the personal danger he is in as landsman, Ted narrates with shifting tenses: "gonna yank me right off" (future); "they stop" (present); "they saw it at the last minute" (past); "was ready to let go" (past). Because the verb "shut down" is ambiguously present or past, and the contraction 's could stand for "is" or "was," the tenses blur even further, giving the impression of a general story that Ted experiences repeatedly and a specific story that Ted repeats in this telling.

Whether specific or general, the story includes a dramatic reenactment of Steve yelling to get attention, which, like arms waving, is a necessity with motor noise. Since Steve is even gentler than his grandfather, Steve yelling is an amusing anomaly to the crew, and even—or maybe especially—the children like to imitate the booming voice he is capable of. In the 2013 fishing season, Steve was again trying to ward off a boat heading for the net, and within moments of it happening, it had become a story. Having been walking down the island's center path while the incident happened, I was made aware of it as soon as I reached the bank and had heard the story told twice before the fishing boat reached shore. In each telling, imitations of Steve's hollering to the interloper to stop, including an exasperated "you idiot!," featured in the story. The name calling is significant to the story because it was extremely unlike Steve's customary gentle behavior. At the same time, in the big picture of stereotypical masculine anger, the language is, well, really tame, and thus amusing for its irony as well as its novelty. As soon as he reached shore, Steve anxiously asked witnesses whether he had spluttered "you idiot" before or after the person stopped, and was visibly relieved to discover that his utterance had occurred while the boat was still headed for the net. While he could forgive himself an outburst in a moment of extreme anxiety for the net and landsman, to insult someone who had heeded him would clearly violate his code of civility, and he would have failed to represent the fishery properly.

Two years later, much to Steve's chagrin, when the crew of the television show *Bizarre Foods with Andrew Zimmern* came to cover shad at the Lewis Fishery, another motorboat headed for the net, threatening to ruin the filming as well as the haul. Steve yells, waves his arms, and in a final act of desperation picks up the silly fiberglass oar used to the steer the boat when it's pulled upriver and slaps the water. The television crew catches the whole incident and goes with the story, blurring the motorboaters' faces as though they were criminals and airing the landsman's and televisions host's laughter. Steve was mortified that his foolish display was aired and became the focus of attention when others discuss the episode. For their part, crew members, family, and friends feel compelled to tell others who have seen the episode that the behavior was out of character. By adding a layer of narration to the media story, people who know Steve clear his name and ratchet up the story's hilarity.

Perhaps the most artful surveillance story concerns a Jet Skier who got caught in the net. Here, Steve's exasperated utterance was clever, self-deprecating, and more controlled, and thus more within his comfort zone. In this story, the stranger, having missed or ignored all of Steve's yelling and arm waving, heads straight for the net, gets the net wrapped around his propeller such that the engine chokes, and is left trapped, bobbing in the water. Waiting for the old Army Corps of Engineers rowboat to make its ponderous way back along the net in its wake must have seemed interminable to the Jet Skier. When Steve finally reaches the man, he asks incredulously, "I'm shouting and waving my arms, telling you to stop; why didn't you stop?" The Jet Skier sheepishly responds, "I didn't see you," to which Steve splutters the story's punchline: "I'm a very large man in a very silly boat; what didn't you see?!" This line always gets a laugh and sometimes leads to discussion of whether the Jet Skier really did not see him or thought he could jump the net in the heat of a daredevil moment. The phrase "large men in silly boats" has become something of a slogan, pinned up at each Shad Festival or other photo display. In its entirety with the ending "what didn't you see?!," the slogan pairs humility with an awareness, even pride, in what the fishery does, although it cannot compete with powerful motorboats and sexy Jet Skis.

This story, like others, reflects the fishery's awareness of others' *lack* of awareness of haul seine fishing, and the need for the fishery to look out for itself through surveillance, surveillance stories, and sometimes calling on the law, as in a case when a boater in the net ended up with points on his license. Told with a more serious tone, in a story from when Fred was captain, a motorboater goes through the net without even slowing down, then speeds off, figuring he can outstrip the fishery crew's oars. However, a couple of men on the crew, aware of the various docking places, jump into one of their trucks and confront the man when he returns to the boat ramp. Some versions of the narrative involve the boater trying to hide from the obviously angry crew, but eventually he realizes that he has no other choice but to return to the ramp where his car and trailer are. In this story, as in Nell Lewis's recounting of Bill Lewis Sr.'s testimony to the federal commission, outsiders are outwitted by the locals they expect they can trick, and knowledge of the river constitutes the locals' advantage.

Knowledge of the river is central to the other type of surveillance story: flooding. In this microlegend type, the extended Lewis family needs to protect the island from the river itself, then pick up the pieces and repair the island afterward, putting it to rights and opening it to the public again to enact the Big Story of sharing the island. In *The Sociology of Community Connections*, John G. Bruhn writes that disasters

> create an urgent need to act against threats to life and/or property and to restore a sense of routine and normality. They happen suddenly with little or no warning and their impact results in the collapse of individual and communal bases of identity. A disaster is not an isolated event, state, or condition, it is a process that sets off a chain of events that triggers further events and responses for years, and even for lifetimes. (2005, 102)

When the Delaware River floods—and typically it does so with just a few hours' warning—Lewis Island is one of the first places hit, being *in* the river. As the island disappears below the water and the bridge becomes impassable, those who identify with Lewis Island and Lambertville can become disoriented with their landmarks obscured, become spiritually displaced when they cannot sit on the benches to watch the river and meditate, and become detached from community when they cannot casually visit a friend on the island.

The repetitive nature of flooding affects people in this area for lifetimes—either one is preparing for the next flood whenever it happens or is still catching up from the last one, or some watermark or missing object reminds one to tell a story of one flood or another. Flood memories begin to run together, and flooding is part of the local worldview. Delaware flooding is a regular enough occurrence that it shapes the built landscape. For example, the house on the island has no basement and the ground floor holds items that can easily be sacrificed, many of them, such as resin lawn chairs, having been salvaged from the trash to begin with. Before each predicted flood, a handful of people—usually Sue and Steve plus members of Pam's and my families—gather in a horrible downpour to hurriedly move valuable items (e.g., the fishing net, the generator, gas grills) to higher ground up on the house's porch and in the crew's cabin.

Fishery family and crew living in Lambertville monitor the height of a rising river, while Steve does the same using reports shared by radio and internet sources. Steve bears in mind his grandfather's traditional measures for confronting desperation (when it's no longer safe to cross the bridge or take a boat across the creek), and the locals' observations sent by phone, text, and e-mail tell him when it's time to get things moved and to lock the gates on the footbridge. The local crew also babysits the boat, every few hours moving its tether up as the river rises so that the boat does not get pulled under the water. Then, as the water level drops again, they move the tether down, so that the enormously heavy boat does not get hung up on land. The biggest point of interest is when the bridge goes out, which indicates how difficult cleanup will be—because of both the work of rebuilding the bridge and the need to ferry people and supplies across Island Creek until the bridge is restored. If a washed-out bridge and the fishing season converge, fishing takes priority over bridge building, meaning that the crew and the catch will be shuttled by rowboat. Disruptions to work patterns then disrupt visiting and storytelling. Against this context of perennial respect for the river's danger and the excitement and tedium of responding to flooding, crew and media tell microlegends of floods.

The history of flooding on the Delaware River provides context for experiences and narratives about them. According to Mary A. Shafer, author of *Devastation on the Delaware: Stories and Images of the Deadly Flood of 1955*, the Flood of '55 is the worst recorded flood in the river's history (2005, 62). The National Oceanic and Atmospheric Administration (NOAA) lists the "Top Ten Highest Historical Crests: Delaware River at New Hope" (2015) in this order: August 1955, October 1903, April 2005, June 2006, March 1936, September 2011, January 1996, May 1942, May 1984, and March 2011. In the narratives told at the fishery today, a nod may be given to the flood of 1903 (which destroyed the last wooden bridge between Lambertville and New Hope), various hurricane names and damages are recalled, and stories of flood cleanups in the twenty-first century have begun to blend together in memory. The flood on the Delaware is the Flood of '55, caused by Hurricanes Connie and Diane, and stories of this flood continue to be passed orally across generations. Throughout the area, one can still see markers of the high-water levels, and the house and crew's cabin on the island serve as measuring sticks, with the water reaching past the sill of the window in the front gable two floors above ground level.[1] Whenever the topic arises, Muriel Meserve usually recalls watching from the family home across Lambert Lane on the mainland as her father and uncle used the fishing boat to retrieve valuable possessions from the house on the island. The image that sticks with her is of her father lying down in the boat so that he could float under the electrical wires and not make contact.

The region and the Lewis family in particular changed with the Flood of '55, and Fred Lewis would track family history by saying that "after the Flood of '55, we bought the house up on Elm Street [several blocks north and inland, where Pam's family now lives], and we moved up there, and then we had to come down here every summer since then" (March 1, 1996). He went on to explain that "this winter [1996] we were going to stay [on the island]; this winter we had heat put in the house," and finally planned to live full-time on the island forty years after the flood. Fred continued:

> But then, in December, Nell had the heart attack and she had open heart surgery, so Muriel said, "You come and stay with me then" [in the house bordering Lambert Lane]. So, we did that. And then with the flood, the bridge went away. We had ... the blizzard. [laughs] ... Couldn't get over there because of the depth of the snow.... Then the floods came along, took the bridge out. And we'll be ready to put it back in a week or so. (March 1, 1996)

This narration links two floods—1955 and 1996—both of which affected the family for months, impacting their living situation and ending with building another footbridge. Fred Lewis's motto, "We've built more bridges than Roebling," underscores the point that responding to floods is a way of life on Lewis Island. Pam Baker recalls hearing the stories of getting off the island just ahead of each flood, giving a particularly dramatic rendering of the family's evacuation before a "freshet":

> That's what they called it where the ice comes down, and those kind of memories of [dramatic voice] taking Don, and getting off the island 'cause, and he was a baby, and the ice coming up, and, you know, right in front of the bridge and those kinds of acts of nature that they just all lived through, in fact. And when I was in seventh or eighth grade, we had one. And I don't think we've had one since. That's where these huge chunks of ice just kind of lay on the island until they melt. And they're amazing, but, yeah, those kinds of things.... But yeah, [dramatic voice] taking Don and running down the island ... (October 4, 2009)

The Don in this story is Donald Lewis, Fred Lewis's nephew, who lived on the island at different times and was a local character into old age. Pam's retelling reflects the perennial panic of evacuating the island nearly annually, and her good humor about the patterned stories passed to her in safety after decades of such close calls.

During my first months of fieldwork and before I ever set foot on Lewis Island or saw a shad haul, the footbridge was rebuilt, and it lasted about eight years, possibly the longest ever for a Lewis Island footbridge. After a couple of harrowing years of moving Shad Fest crowds over to the island five to ten at a time in each direction while a crack in a joist gaped, the family took down the first-ever worn-out bridge in 2004, which ironically turned into another flood microlegend. The setting is a significant time for the fishery: Fred Lewis had passed away during the shad season, and then the fishery had taken part in the Smithsonian Folklife Festival's exhibit of mid-Atlantic maritime traditions in the summer. Ownership passed to Nell Lewis, and leadership rested more completely on Steve's shoulders. One beautiful September weekend between the fishing and snowing seasons, a loose conglomeration of family, crew, and friends gathered to tear down the bridge and put up a new one. The group's children were playing within a fenced enclosure up near the house, the bridge was down, and about $1,500 worth of new wood—a rarity for Lewis Island bridges—was being readied for assembly into new trestles. When Steve looked for a rock he had placed at the river's edge and couldn't find it, he thought at first that someone had thrown the rock away. Taking a second look, he suspected that the river was rising fast, and so he walked to the edge and measured the height with his arm twice over a short period of time. He calculated that it was rising about a foot an hour—an extremely rapid rise—and, although it was a beautiful day with no flood warning, started safeguarding the island.

Quickly Steve called for the work party to switch to new priorities: get the children off the island, tie down the wood on the island and the mainland, and gather up all the tools and other things usually moved to higher ground during flooding. The two four-year-olds in the group were carried down the island at a rapid pace, much like baby Don had been sixty years and three generations earlier. Crew member Dan Tuft was asked to tie the new lumber together, then tether the bundle to a large tree on the island. Although Dan has regularly helped move valuable materials to higher ground in expectation of flooding, this responsibility probably weighed the most heavily on him. Knowing only about two knots, Dan made up a knot in these moments of panic and was immensely relieved to see the wood still there a few days later. Dan's role in securing the expensive wood is given a heroic nod much like Steve's role in detecting the coming flood, but where Steve's contribution shows the triumph of traditional knowledge (of the river), Dan's contribution reflects the island and fishery traditions of helping as best you can, making it up as you go along, and hoping it works out OK.

The last retold feature of the 2004 flood, in contrast, is about lack of knowledge. When the flooding became obvious to others in town, people unrelated to the fishery noticed that the footbridge to Lewis Island was gone and assumed that it had gone down in the flood. They might have thought that both ends and all the trestles had come loose, or, more likely, they had never before noticed that when the bridge goes down it stays floating in place until it is rebuilt and so did not see that it was different this time, with no bridge parts in sight. As Lambertvillians began to tell the new flood stories to each other, those related to the fishery told the others that, no, the bridge had actually been taken down intentionally beforehand—and ironically, the flooding could have saved them some of the deconstruction labor, if they had known it was coming. This part of the story, discovered in conversation as fishery people revealed it to others, was then integrated into the story later as evidence of the confusion that attends floods, as people look for landmarks in the disruption. At the same time, the bridge misunderstanding fits another aspect of the fishery in local life: it is the thing that is ever part of the environment that many people do not see completely.

The 2004, 2005, and 2006 floods together create a greater regional narrative that combines elements of the known and the unknown as experts, media, and flood victims debated the degree to which damage could have been prevented or could be prevented in the future. While the weather was and had been gorgeous that September 2004 day, there had been heavy rains upriver that caused the flood. At the same time, reservoirs in New York State continued to release water into the Delaware. Two more such floods in 2005 and 2006 exceeded the flood records of the 2004 flood to take third and fourth places on NOAA's "Top Ten" list (2015), on which the 2004 flood does not appear. After the series of three, however, analysis by collaborators from the United States Geological Survey, Delaware River Basin Commission, US Army Corps of Engineers, and NOAA (Goode et al. 2010) determined that managing the reservoirs differently would not have *prevented* the 2004 flood, but they also laid plans for more watershed management cooperation in the future, with New York state authorities agreeing to do their part (Duffy 2011). Much of the controversy, played out on blogs and in casual conversation, seems to be about how managing waters (e.g., keeping reservoirs lower in non-flood times and not releasing waters during times of heavy rain) might have saved some degree of damage. The savings would be terribly important to people affected by flooding, who would nevertheless still experience some damage, but the gain might seem a pittance to those managing the water. Matching the ethos of Lewis Island, this is essentially a story about civility and neighborliness that flips the story of restoring the shad through environmental management: what folks do upstream (emptying their reservoirs during heavy rains) can greatly affect those downstream (flooding) just as what folks do downstream (pollution) can greatly affect those upstream (no shad able to penetrate a pollution block). In both cases, people need to consider the needs of others when acting.

Other microlegends of the 2004 flood highlight stewardship in their concentration on restoring the island so that it could be reopened to the public, stories that melded with those of other homeowners in Lambertville, and up and down the Delaware. On the island, the flood left behind three inches of mud in the house, pulled out the utility pole bringing electricity to the island, and left a ribbon of debris in the trees high above one's head as well as underfoot. For several weekends, those of us living on the hills in Lambertville would have a role helping one family or another downtown, and dumpsters sat all over the low-lying parts of town. In a flood, each room and space becomes a collage of someone's life, sometimes saved above the brown water, sometimes churned in the mud, sometimes tossed in the dumpster. I realized just how symbolic the dumpsters had become when my four-year-old, Adelaide, looked at our next-door neighbor's construction dumpster up on our hill and asked, "Mommy, how come the flood came to Jud's house, but not ours?" Before she manufactured an interpretation along the lines of the Passover story or feared losing her hilltop home in the next flood, I explained that not all dumpsters mean flooding, and I began to see how the flooding affected the kids' worldview. Fishery kids had long been aware of the dangers of water, having to wear life vests at all times while fishing, but they now saw that it presented other kinds of danger. Flood-induced erosion had stripped away the soil from exposed rocks and roots, causing trip hazards, while the river brought broken glass, hypodermic needles, and other sharp objects. During the work parties, we would wear gloves and boots and before lunch conduct elaborate cleansing rituals using what seemed like gallons of hand sanitizer.

When the bridge was finally rebuilt, it did not even make it through a fishing season: the flood of early April 2005 washed it out again, and then the June 2006 flood washed out the next footbridge. Each time brought a little less damage, because with each flood people had less to lose, having thrown it out the flood before, and had become more practiced in flood preparation. The house on the island no longer has carpeting on the wood floors. After repeated floods, some local people sold their homes, and others raised their houses up a story. Cleanup could become slower as one layer of damage went overtop the last, or it could become faster as people prepared better and loosened their standards of avoiding contact with mud and water. While the news helicopters overhead might have felt exciting the first time, it became unsettling day after day in the next floods. Moreover, the catastrophes of the tsunami in the Indian Ocean (2004), Hurricane Katrina (2005), and Hurricane Sandy (2012) put the Delaware floods into perspective, and regional residents responded with understanding and assistance while still recovering from their own repeated disasters.

Generally, stories of cleanup after the second and third floods blend together in a same-old-same-old description rather than distinct narratives. However, Keziah tells this story of the crew turning the flood cleanup into fun when she was about nine:

> Well, it went all the way up to Mrs. Meserve's house [meaning the river rose from its banks and crossed the island, the creek, a parking lot, and Lambert Lane], and all this like sewage and stuff got all gross, and it filled the whole parking lot. So, we were all in our boots, and we were cleaning out the bridge—and it's like two foot high of mud or something; I'm not sure. And before I came—I didn't even know it—they took everyone's pocket change. It was like five bucks' worth or something, and they said whoever falls in first or is about to fall in and gets stuck, can get it first, and I got stuck first . . . and I had . . . one shoe got completely stuck in there, and then my other foot I had to step in the mud, and then Ted had to come and pick me out of it. It was big . . . because I got the money. (October 3, 2009)

Here, the child's experience is turned from danger and drudgery into protection and reward. Despite the "grossness" of the flood cleanup, there is a sense of comradery and a growing attachment to the island with each flood. Keziah says that she often tells flood stories to island visitors or friends: "People are like, 'oh, you live on a hill, you're safe.' And I say, yeah, but the fishery isn't" (October 3, 2009); clearly she feels a connection with the island that includes stewardship.

Pam connects the flood stories she heard growing up to a sense of well-being as well as connection to the island: "Even though the Flood of '55 was awful, they came back from it. So they have these great memories of it. And even in hardship, it's 'oh, we got through this.' You know, something to bolster yourself up, and I think that's part of it" (October 4, 2009). Clearly, the flood and other surveillance stories are moral narratives of fortitude, resilience, and civil community, while the wonder stories teach about human nature and the power and mystery of the river. The microlegends are also entertaining to hear and to tell. They connect people as does singing a song's chorus, for when told within the fishery's small community, these stories do not exchange information as much as remind people of lessons and connections.

There is a continuum of inclusion with microlegends. When people ask, "Were you there when . . . ?" to start a story, this can be characterized as a tool of exclusion, as in "ha, ha, I was there, but you missed it." However, as the interviews express regret at not witnessing some of the occurrences, they also tell the stories. Often a phrase like "I was not there that day, but ..." will serve as a disclaimer and preserve accuracy, but then the narrators proceed, claiming the story as their story in the sense that it is the fishery's story, the island's story, or a story of the Lambertville–New Hope area. Fishery kids may tell stories or serve as the primary audience for stories, with new visitors listening and receiving the story as well. As children, Sarah Baker and Adelaide Groth-Tuft began directing the storytelling at quite a young age by asking for specific microlegends that they clearly already knew. Visitors to the island may also incorporate themselves into the story's community by mentioning how they found out about a particular happening, while crew and family members have been over the story so many times or are in such regular contact during the fishing season that they seem to learn by narrative osmosis.

The issues of belonging here get blurry, as characters, witnesses, tellers, and audiences all become part of the process of turning occurrences into microlegends. Curiously, when crew members share microlegends with reporters, these stories rarely if ever end up in print, perhaps because feature reporters have been asked to cover the Big Stories, perhaps because legends do not function well as "news" stories and cannot be told economically to strangers, or perhaps because the journalistic style of making the teller invisible renders the connective storytelling nonfunctional. Barbara Johnstone describes the process of changing personal experience narratives into communal stories this way:

> As they are told and retold, narratives of personal experience become more and more detached from the real world. The general point of the story becomes increasingly important, and the details decreasingly important. A story originally about something that really happened to an actual named person, at a particular place and time, gradually becomes a tale which illustrates the sorts of things that can happen to anyone, anywhere, and at anytime. (1990, 133)

This may be why fishery family and crew members ask each other—and sometimes ask frequent island visitors and customers, too—"Were you there the night when ... ?" Any of them could have been there, and many people can serve as a storyteller or an audience member, able to add a forgotten detail or chime in on the theme as in a chorus.

The footbridge, looking from Lewis Island toward the locking gate on the mainland side and the Lewis family's shuttered home on Coryell Street, 2008. Photo by Adelaide Groth-Tuft.

The crew's cabin and drying nets with the center path leading to the house, 2017. The far bank is New Hope, Pennsylvania. Photo by Lita Sands.

At the start of a night's fishing, Dan Tuft, Tim Genthner, and John Baker change into boots in the crew's cabin, 2008. Photo by Frank Jacobs III.

Chatting and mending net while loading the net on the boat, 2004. Left to right: John Baker, Steve Meserve, Jack Marriott, Adelaide Groth-Tuft, Keziah Groth-Tuft, Jessie MacGregor, and Tim Genthner. Photo by Randy Carone.

The crew drags the rope up the outside path, past the house and steps, 2004. Sue Meserve steers the boat and supervises the kids. The Route 202 bridge stands in the background. Photo by Randy Carone.

John Baker, Sarah Baker, and Steve Meserve pull the boat up the outer path past the house while Charlie Groth steers. The New Hope-Lambertville Bridge stands in the background. Photo by Frank Jacobs III.

The brail holding the net's land end falls off the back of the boat as Steve Meserve, Andrew Baker, and Rich Pfaff start the haul, 2017. New Hope, Pennsylvania, can be seen in the background. Photo by Lita Sands.

Ted Kroemmelbein holds land using a smalick (leather harness) while walking down the outer path built by Theodore "Uncle Dory" Lewis. The New Hope-Lambertville Bridge can be seen in the background. Photo by Paul Savage/*Times of Trenton*.

When they get close to the bank, the crew uses a pole in addition to rowing, 1937. Built specifically for shad fishing, the old boats had room for three rowers, a poler, and the loaded net. Photo from the Lewis family collection.

Steve Meserve poles while Dan Tuft (wearing hat) and Tim Genthner row toward home, 2008. Photo by Frank Jacobs III.

Sue Meserve, Keziah Groth-Tuft, and Charlie Groth (left to right) catch the boat while Steve Meserve prepares to hand off the sea end of the net and Dan Tuft (left) and Tim Genthner (right) finish rowing, 2004. Photo by Randy Carone.

John Baker brings the net's sea end down to the point of the island, while Sue Meserve rows the boat around to the creek side of the island, 2004. In the background, Charlie Groth and Adelaide Groth-Tuft walk to the point to pull in the net. Photo by Randy Carone.

In lower waters, the crew pulls in on the side, 1963. Donald Lewis (far left) and David Meserve pull the cork line while Robert Johnston (left) and Frannie Williamson (right) pull the lead line. Frannie's sons Steve and Skip look on, as do Fred Lewis (by the sawhorses) and his sister Eek (Edith). The stone and cement apron, built by Theodore Lewis, keep Lewis Fishery shad clean as they are pulled in. Photo from the Lewis family collection.

Adelaide Groth-Tuft, Charlie Groth, Pam Baker, and Dan Tuft pull the cork line, 2011. Tim Genthner (yellow cap) pulls first leads, while Ken Bacorn coils net. Ted Kroemmelbein pulls the land end cork line at the far right. Photo by Muriel Meserve.

The crew pulls in the net over burlap to avoid catching it on cracked cement, 2017. On the sea end cork line, (left to right) Brian Nagengast, Charlie Groth, Peter Hollis, and Andrew Baker. Adelaide Groth-Tuft pulls first leads while Rich Pfaff coils. On the land end, Ted Kroemmelbein and Aiden Nagengast (looking at camera) pull the cork line, with Jim Coll and Steve Meserve (far right) on the lead line. Photo by Lita Sands.

It's not until the final moments of the haul that the crew can tell what sort of catch they have, 2008. Crew (left to right): Sue Meserve, Keziah Groth-Tuft, John Baker, Steve Meserve, Dan Tuft (pulling leads on right), Charlie Groth (foreground), Adelaide Groth-Tuft (wearing life vest), Sydney Osler. The background shows a common crowd for a weekday evening, including long-term customers Dr. Dutta (cardigan), Morgan Van Hise (dark cap), and Ed Padilla (center back), as well as future crew members Robert (backwards cap) and Matt Eick (light sweatshirt and cap). Photo by Frank Jacobs III.

A typically small Saturday morning catch, 1996. Standing: Johnny Isler and Norman Hartpence (with cap). Crouching (left to right): Sue Meserve, John Baker, and Steve Meserve Photo by Charlie Groth.

After "sexing" the fish by sorting roes (females) into the left-hand bin and bucks (males) into the right-hand bin, Sue Meserve asks customers to wait a moment while she decides how to divide up the haul. In the background, Charlie Groth weighs a shad and Pam Baker records data. Photo by Stephen P. Harris (www.sph-photo.com).

William Lewis Sr. mends net with his dogs, Blackie (the German shepherd) and Carlo. Photo from the Lewis family collection.

Fred Lewis rests in the boat before a haul, ca. 1970. The homemade oar (a two-by-four with a rounded hand grip) can be seen. Photo from the Lewis family collection.

Steve Meserve mends net, 2008. Photo by Frank Jacobs III.

Three generations of Lewis women at the family home on Elm Street, mid-1990s. Left to right: Sue Garczynski, Nell Lewis, Muriel Meserve, and Pam Baker. Photo from the Lewis family collection.

At the end of a haul in 1937, with the crew's cabin and house visible in the background. Crew in front, left to right: William Lewis (far left), Elmer Stout and William Stout (holding fish), Red Bair (holding line), Theodore Lewis (holding line and fish), and Lyman Stout (far right, front). Among the onlookers on the bank is Eek (Edith) Lewis, William Lewis's daughter. Photo from the Lewis family collection.

The crew poses for a picture in front of the cabin during the Shad Fest, 2004. Several crew members served as pallbearers for Fred Lewis's funeral that morning, and their black ribbon armbands can be seen, signifying that the haul was performed as a memorial haul to honor Fred. Front row, left to right: Tim Genthner, Charlie Groth, Jack Marriott, Steve Meserve, Dan Tuft, and Sue Meserve; back row, left to right: Andrew Baker, John Baker, Sarah Baker, Keziah Groth-Tuft, Adelaide Groth-Tuft, Ted Kroemmelbein, and Pam Baker. Photographer unknown.

In winter 1996 flooding, water reached a few feet into the area underneath the crew's cabin. Photo by Steve Meserve.

Lambertvillians enjoy swimming from Lewis Island and sunbathing on what has been called Lambertville Beach. Visitors could buy refreshments under the cabin. Photo courtesy of Jeff Kline and James Mastrich, taken from *Images of America: Lambertville and New Hope* by James Mastrich, Yvonne Warren, and George Kline.

FIVE

"It's Like I Said to So-and-So": Everyday Storying

> With fishing . . . even if it is just three months, it's that three months
> of connecting with each other on a daily basis—you know, how was
> your day? You know, actually tell me what you're feeling.
> —**Sarah Baker,** fifth generation of the Lewis Fishery (September 13, 2012)

I first heard the common phrase, "It's like I said to so-and-so," in the mid-1990s. In everyday conversations, my friend Deb would share something she had already discussed with someone else, referring back to the earlier conversation. For example, she might say something like, "It's like I told my sister: if I had my way, . . ." and then she would go on to repeat the material to me. At first I was disoriented, trying to figure out how her sister was related to our conversation. Eventually I realized that this practice is something like the flipside of what Richard Bauman discovered when he noticed men telling stories, then attributing them to an earlier teller the stories "belonged" to. Bauman interprets this practice as preserving the "personal involvement" "by placing the secondary narrator in a chain of transmission that linked him through his source to the event itself" (1972, 335). Similarly, the "it's like I told so-and-so" phrases reach back in time to incorporate the current listener into an ongoing conversation, perhaps building the connection between the first and second audiences. When Deb's conversation topic was academic, the linguistic device gave credit to the first listener for molding positions developed in academic discourse. It was also a neat solution to the linguistic and microethical problem of having the same conversation with different people, being aware that one had practiced and developed an argument, and not wanting to sound rehearsed. The forthright admission of repeating material acknowledges that the material is not new, and at the same time the "like" releases one from quoting oneself verbatim. In the years since I first heard it, "it's like I told so-and-so" has become a common expression in the Delaware Valley. At the fishery, the phrase not only brings the listener into a community discourse of everyday lives and connections, a practice I call "everyday storying," but also calls attention to the collective development of experience and story in a near-ritual practice.

Everyday storying is telling stories in a quotidian sense—everyday stories, everyday performances, told almost literally every day. Over thirteen years into my study of the fishery, I began to reinterview fishery family and crew members with the concept of narrative firmly at center. Wanting to cast the net wide and get my interviewees to consider personal experience narratives (PENs) as well as the kinds of discrete stories they were accustomed to seeing at staged storytelling sessions and in published collections, I initially gave a rather vague description of what I was looking for. I prompted Sue and Steve Meserve this way: "When I say 'story,' I don't mean simply that story that's polished and finished that sometimes we think about, but also any narrative, any recounting of events, recounting of things that happened, or fictional recounting. It's quite wide, the idea of 'story'" (October 9, 2009). As the interview continued, a wide range of narratives emerged that could be typed and interpreted later. Looking back, I see that—much the way composition instructors ask their students to develop thinking *through* writing—we were shaping concepts of narrative together based on conversations about observations and experience, rather than going looking for well-known types with a butterfly net and field guide and thus seeing only what we searched for. As a result, my concept of everyday storying emphasizes the fluidity of storytelling as a process *and* as a product.

The folklorist will likely recall Richard Bauman's article "The La Have Island General Store: Sociability and Verbal Art in a Nova Scotia Community" (1972), in which he analyzes talk for talk's sake among men doing their shopping. Bauman divides the content roughly into three types: news, yarns, and arguments (1972, 337), and these three verbal arts subgenres can be found on Lewis Island, although the nature of each reflects the concerns of the community there. On Lewis Island, most everyday stories are PENs or friend-of-a-friend (FOAF) stories, in which the teller shares his or her own experience or shares secondhand knowledge, but it is also quite common for people to step aside further to integrate into the conversation stories from history, current events, and traditional lore. Thus, "news" on Lewis Island blurs the personal, community, national, and global levels. The nearest thing to Bauman's "arguments" (1972, 334) on Lewis Island take on the academic sense of the word somewhat, usually tempering emotion and foregrounding reflection and analysis. Unlike "yarns," which Bauman says must "transcend common knowledge, experience, or expectation" to get "foregrounded" (1972, 334), stories do not have to compete to meet standards at Lewis Island. Unusual FOAFs and fictional, even literary, tales may get interspersed with mundane

PENs as the conversational topic dictates, creating conversations that are even "story dominated" (Hufford 1992).

On occasion, the Big Stories, character anecdotes about captains, and microlegends will be integrated with everyday storying. Resembling microlegends in that their scope and import decrease as they move along the continuum away from myth, everyday stories are likely to involve the storytellers as characters. On the other hand, everyday stories and microlegends differ in their sense of discreteness, repeatability, and import. Everyday stories, such as the "how was your day" variety, are more personal, less important to the world at large, and perhaps even less important in content to the tellers and listeners than some other type of story; they might not bear repeating. The story may not have a clear beginning or end, and the lines between storytelling and conversation blur easily. On the continuum between art and communication, everyday storying favors *commune-ication*—building relationships.

Everyday storying crosses with the local verbal genre of "chitchat," a term that for decades titled a column in the local newspaper, the *Lambertville Beacon*. It has its parallel in the idea of "craic" in Northern Ireland (see Cashman 2008), and *The Concise Scots Dictionary* (Robinson 1985) uses various spellings for verbs (e.g., "boast, brag"; "talk, converse, gossip") and nouns (e.g., "a talk, gossip, conversation"; "a story, talk") that refer to the Scottish equivalent. Not only does a continuum between discrete story and conversation exist, but one term—craic, chitchat, everyday storying—can be used for specific narratives as well as the verbal activity that blends and intersperses utterances from different parts of the story—conversation continuum. Thus, the verbal construct story*ing* bridges content and (inter)action.

"The daily grind," PENs about everyday life, forms the mainstay of everyday storying by subgenre. "How was your day?" conversations with various tellers and listeners, whether crew members, family members, friends, customers, or other visitors, recycle and repeat over the course of a night's fishing. Sarah Baker, a young adult, describes this practice as a valued part of the fishing season that contrasts with the brief, more superficial interactions she has on social media: "With fishing . . . even if it is just three months, it's that three months of connecting with each other on a daily basis—you know, how was your day? You know, actually *tell me* what you're *feeling*" (September 13, 2012). For Sarah, the details of the daily narratives are infused with emotion and connection, even if the content might be mundane. Stories and even ongoing sagas from family life, school, and work dominate generations of conversations on Lewis Island. Diane Kroemmelbein describes such conversations going back to the 1960s and 1970s: "I remember when I was a kid, Fred and my dad would talk about—and I don't remember any of the stories,

I just remember they would talk about some of the funny things that would happen at work, or something, you know. Not a whole lot, but often" (June 15, 2013). Her father, Johnny Isler, and fishery captain Fred Lewis worked at the Union Mill during the day in different capacities and then might discuss mill happenings at the fishery, where they worked more closely together. Diane continues: "They knew the same people then, and just stuff that would happen, you know, like if somebody cut their finger off, or, you know, just stuff" (June 15, 2013). Here, Diane groups a large range of topics from the mundane and amusing events she can't remember to a life-changing injury.

Similarly, a few years earlier Keziah reflected on her conversations with Sarah, just two years apart in age: "Sarah and I, we talk about school more because we have some friends or acquaintances, in common? [Charlie Groth: Mm-hm.] And so, our stories will more be about our—we talk a lot about *our* days and stuff, more like *friends* than like I do with adults" (October 3, 2009). Because of shared social networks, Sarah's and Keziah's conversations resemble those of Johnny Isler and Fred Lewis, three generations earlier. Keziah points out here that she is also included in conversations with the adults who do not share her school-based social networks, and this is different because of both the age differences and the degree of separation between audiences and subjects. In the current crew, the kids and the adults report on daily happenings occurring at different schools and workplaces, whose "characters" have become familiar to audience members at the fishery with each saga's installment. Listeners know the rhythm or "time setting" of each school or workplace—the run of a show, a semester or marking period, a sports season, a response to a disaster, prom and concert schedules, construction seasons—and follow the particulars in their friends' lives.

Sagas about creative projects form one subgenre in "how was your day" conversations, bridging work and home worlds. For Sue Meserve, these projects are part of her paid jobs in contract work and theater. For others, however, creative projects tend to happen only on the home front, and the crew actively requests and listens to stories about a latest home improvement project, what Tim or Ted is working on in his backyard pond, or one of the kids' art projects. Such talk usually has a very limited interest for audiences; that is, there's only one guy at someone's workplace who will tolerate another story about removing dead limbs from a buttonwood tree. However, crew members of all ages will listen carefully and help problem-solve, in part because of the hands-on nature of the group. At times, crew members will answer the call for help at each other's houses, but more often collaborative building or maintenance work happens at the fishery; creative work at other locations is re-created and shared mostly through narrative.

Moving out from PENs, conversations may quickly shift to include general topics, such as fishing, national or global news, and history. Contrary to what people usually expect, only one or two of the crew members like to talk about line fishing. Most of the crew members don't do line fishing, so everyday storying about line fishing is a likely topic for conversations just with island visitors (particularly males), many of whom come to the island primarily because of their interest in fishing (not in the Delaware River, Lambertville traditions, or the environment). Thus, line fishing is something of a "general" topic for stories that cross the crew/not-crew line, as do history and current events. In an October 3, 2009, interview, Steve and Sue Meserve discuss the mix of topics at various times during a haul:

> SUE: Well, it'll run the gamut, I mean, you're still going to hear about people's day, or the state of the union. Ted's good on that one too.
> STEVE: Yeah, he is good on state of the union.
> SUE: His stories will gravitate toward politics and his disdain for politicians.
> STEVE: Organized government . . .
> SUE: Yeah . . .

Here, "good" means engaged, knowledgeable, and happy. Sue and Steve agree that Ted will have a lot to say about the topic of politics and will keep the conversation going on this topic. The comment reflects the fact that certain topics can come to characterize particular group members. For example, Sue is associated with stories about history, and Ted, myself, and others are viewed as likely to join in on that topic (Keziah Groth-Tuft, October 3, 2009). It is also important to note that while the daily grind may offer some points of entry into the crew's conversation for island visitors to join in, the broader history and current events topics also allow others to contribute. However, when the audience broadens, topics are chosen more carefully to limit controversy and promote civility. For example, a recent tragedy will be a more likely topic than politics when the group includes customers and visitors.

Moving in the other direction from national and global current events and history, "how's your day" reporting might turn inward to troubles talk. Immediately after mentioning Ted's penchant for talking politics, Sue Meserve adds:

> But it gets more personal. I mean, people'll sidle up to you and, you know, like, Tim will start with me, "How's your knees? How's your knees been doing," or "What's going on with your heart these days? That all taken care of or what?" Or you'll be next to somebody and you'll remember something they said last week, and you know, whatever happened to that? (October 3, 2009)

Just as people will ask for the continued story of a project, they will keep tabs on each other's personal concerns. Sue mentions health issues here, but such continued stories might also include family problems or problems at work. Here we see the anatomy of a community, as scholar John G. Bruhn describes communities in *The Sociology of Community Connections*: "A cohesive community is also a supportive community. Members help each other with unconditional acts of sharing as they experience the good and bad effects of life events. The community itself is a safety net" (2005, 206). During the fishing season, crew members and island neighbors will be well acquainted with each other's concerns and respond with a home-cooked meal or appearance at a funeral. Off-season, the communication lines rely on telephones and e-mail, wider grapevines, and the local newspapers, but Lewis Island community members respond with interest and acts of caring nevertheless.

At the start of each fishing season, there's a certain amount of "catching-up" storying as many crew members and customers do not see each other from June through February but for the occasional chance meeting on the street or in the grocery store. Outside of crew-crew and crew-customer updates, however, "catching up" accounts for a large portion of the everyday storying between crew and family on the one side and visitors on the other. These visitors may be connected to crew members solely through the fishery, or through other personal or community links.

Typically, the visitors fall into two categories: those interested primarily in the fishing and those interested primarily in the Lewis family. The everyday, catching-up storying on these topics often falls along gender lines: women mainly seek out the information on the family from the women, and men mainly seek out information on the fishing from the men. This pattern reflects in part the traditional norm of same-sex friendship affiliations, but it also reflects traditional gender roles. Meanwhile, the women's work of exchanging catching-up stories on family is consistent with studies on the characteristics of women's conversation work, including "rapport talk" (Tannen 1990). Although individuals will trade the quotidian "how's your day" stories regardless of gender, once the catching up goes further—either over a longer period of time or a wider scope of family—females become overrepresented among the tellers and listeners. Muriel Meserve explains the mutual relationships established through women's catching-up storying about families, and her mother, Nell Lewis, agrees:

> We've always been the one who kept track of all these families, and they would keep track with you [to Nell] on how everybody was doing. I mean, that's how we became friends with the Hentschels, right? [Nell: right.] Because they would

come up [from the Philadelphia area], and, yeah, they'd talk to Dad about the fish, but they would, you know, do all the family chat with Mom, and then started coming to all the family gatherings and things, and, if I'm over there ... they wouldn't ask me too much about the fish, but a lot about how everybody was doing, how all the rest of the family who wasn't there because they get to know everybody, and other members of the crew that weren't with it anymore too; they would also check on *them*. (October 5, 2009)

One or two generations later, Steve Meserve concurs that while people will ask him about his family among other more fishing-related topics, visitors wanting family news will often approach his sister Pam instead (October 3, 2009).

While the traditional gender roles are clearly at work in conversations about fishing versus family, this process also reflects the fact that this family's particular circumstances have led to the women's association with the family line because of appearance and longevity. Pam Baker explains it this way in a conversation with me:

They'll recognize me, because I look like my mother, and so they'll automatically recognize [laughs] [Charlie Groth:—and your aunt—], and my—we all look alike. Or they'll see Sarah and say, "oh, Pam must be your mother, Muriel must be," you know, "you must be related to which"—and it's true, we do all look alike. And with Steve it's a little harder, because you don't—he looks like Cliff [her mother's brother, Fred Lewis's son]. He looks a *lot* like Cliff, and Cliff only lived to be thirty-three, so [Charlie Groth: so few people knew him], so, right, so he doesn't have that longevity. So with me it's more of a "oh, I can ask you" where they might not be positive of the [men's link to the family]. (October 3, 2009)

Of course, one should also note that it's more likely that a female crew member will also be a family member, as men have been more likely than women to volunteer for this traditionally masculine activity. Pam immediately, however, moves back to the more socially gendered explanation:

You know, they know [Steve's] part of the [family], but with, I think, too, females, [speaking in a comic tone] "they know everything. I can just get to the source." The wife will come and talk to me, and then the husband will go and talk about fish.... And that's true. That's funny. (October 3, 2009)

Pam admits, "Well, you know how we [the Lewis women] are: we pretty much tell everything; there's not much we don't tell. Sometimes we try to maybe sugarcoat it a little bit, you know, as [relatives] become infirm. And it's people who grew up with us, or, again, there's older people who, for whatever reason, are in town for a wedding, a funeral, a this, a that, a reunion" (October 3, 2009). This drive for information quickly moves into a give-and-take:

> So they'll want to catch up and they'll want to see, and then they'll share a little bit about, "Well, when I was" you know, "in school," or this or that, but they'll want to catch up and make sure if Grammy's alright or how is she and whatnot, and then you tell the sugar part of it, because she is doing well. And then, same with Mur . . . they'll ask those things and how everybody's doing. And then they'll share family: "I now have however many grandchildren" or all that, but they'll go back to when they used to come down. (October 3, 2009)

Thus, the drive for information about fishery family and the connection to fishing itself become interlinked as people cover both topics, then connect them through their own experience of "coming down," which is a characteristic and highly localized abbreviation for coming down to the riverside and onto the island, especially to the fishery during fishing season. As Pam's description attests, "catching-up" stories quickly turn to reminiscing, and her brother Steve buttresses this point when he describes everyday storying with locals: "They'll want to reminisce too about, you know . . . do you remember *when*, or when your granddad did *this*, you know" (October 3, 2009), the stories being similar and yet too varied for him to pick out particular examples.

Crew member Dan Tuft describes Steve Meserve's own reminiscing at the fishery about his grandfather, whose work and identity as the fishery head Steve has assumed: "Occasionally Steve will tell a story about his grandpa, or he'll say, while we're doing a haul, 'oh, it's a good thing that Grampy isn't watching,' or 'if we were doing this, he'd be saying something to us' or something like that" (October 3, 2009). This sort of story imparts Fred Lewis's standards and knowledge, augmenting the "WWFD" (What Would Fred Do) pattern discussed in chapter 3. Dan goes on to say that Steve would similarly reflect on how certain tools, materials (e.g., netting, rope), or related processes came to be established during his grandfather's time. This narrative practice passes traditional knowledge to a long-term crew member about parts of the operation that someone needs to know about but that are discussed less often than the techniques of the daily haul. Delivered in short pieces, these mini-narratives build a longer history of the fishery's material culture.

As one of the crew members on the rotation between rowing and first leads (the two most physically demanding jobs), Dan is also privy to Steve's reminiscences about the work styles of crew members Steve has worked with over the past four decades. These narratives may arise when discussing a particular aspect of the work (e.g., rowing rhythm, when to pull the pin that attaches the net to the boat), but almost certainly will come up when a former crew member visits the island to watch a haul. Dan recalls,

> Well, we would occasionally go through who used to be on the crew on different occasions, either an ex–crew member comes over, 'cause they were too old to do it now, like Sparky would come over, or when Johnny [Isler] was alive and he would come back over, and then there'd usually be a recounting of who was actually on the crew in days gone by, and sometimes Steve would talk about when they, I think Steve was around when they did three rowers on occasion, 'cause he would describe who all was on the boat, and I don't remember any specific stories, but of how they'd be recounting who would do what, like Sparky would be the landsman, and Johnny would be rowing with, sometimes it would be with his grandpa, and sometimes somebody else? Maybe Steve. (October 3, 2009)

Here, Dan describes and enacts a ritualistic recounting of the crew membership—people who are also story characters. This cast call resembles the "Storyrealm" activity Mary Hufford explores in her study of Jersey Pinelands fox hunting:

> Through stories told in the course of a single conversation hunters can explore a realm spanning sixty or seventy years as they draw occasionally on received memories from the previous generation, leap backward and forward in time, and allow older relatives and friends who have died to speak as cohabitants of this realm. (1992, 174)

Hufford argues that people and experiences spanning multiple lifetimes are invoked through conversation and in a sense summoned into the room with the speakers, some from beyond the grave. At the fishery, names invoke thoughts about men who have died or simply retired. In these conversations, Steve, Muriel, and Pam will likely be the ones to represent the views and experiences of Fred Lewis as well as those of his father, Bill Lewis Sr., which have been received by Muriel directly or received by Steve and Pam indirectly through Fred or Muriel. At the turn of the millennium, Johnny Isler would come back as one of the visiting former crew members who sparked these reminiscence sessions. Since Isler's passing, his son-in-law, Ted Kroemmelbein, might represent his memories during another former crew member's visit, and a visit from Diane Kroemmelbein (Johnny's daughter) might have the same effect of setting off a reminiscence session. Diane's connection with the fishery led to her husband's connection, just as the husbands of Muriel, Pam, and myself, respectively, all became crew members. Her memories begin in her childhood through her father's involvement, and so her visit summons Johnny Isler into the fishery's version of Storyrealm, as if Johnny himself were visiting. At the fishery, however, it is not just the individuals who are recalled through reminiscence narratives but also the different crews' collective identities. The former crew, with its particular combination of quirks and skills, is reconstituted through reminiscence.

Reminiscing at the Lewis Fishery during fishing season need not always be directly about the fishing itself, however. For example, Pam Baker talks of the friendship between Ed and Betty Hentschel (mentioned above by Muriel Meserve) and Fred and Nell Lewis as one that grew up around reminiscing: "They would come whenever they could. And it just turned into such a nice relationship, because it was harder for my grandparents, of course, to find an older couple to talk about their history together, and they just had similar interests" (October 3, 2009). Reminiscence about earlier eras and life stages extends from Fred and Nell Lewis's generation to the fifty- and sixty-somethings as well. In the few years when Jack Marriott was a regular part of the crew and when Pete, another of Steve's childhood neighbors and friends, would stop by to help pull in the net, reminiscences of the people and goings-on at their end of Coryell Street in the 1960s and 1970s became a regular topic (Dan Tuft, October 3, 2009). Keziah says of Jack Marriott, "He and Steve talked a lot about their childhood, because they grew up together, so they would like talk about like one time Jack like went through the window, and stuff like that" (October 3, 2009). She describes these sorts of reminiscences as being "funny" rather than cautionary and reminisced herself in her interview about how much the younger fishery kids enjoyed these stories. Clearly, the fishery crew creates stories on an ongoing, emergent basis, an activity that will be discussed further in the following chapter on processional storytelling.

Although so much of the narrative system has serious underpinnings, on the whole the island is a place for laughter, including good-natured teasing, such as the "hassle Steve" story discussed in chapter 3. Dan Tuft confesses: "We usually razz [Steve] about talking to the press. Tim and I do," "about being famous" (October 3, 2009). On many occasions, I have also observed the men on the crew, with exaggerated deference, kidding Steve about his responsibility to decide who does the hardest jobs, and may at the same time tease each other about trying to get out of work. Both types of teasing point to Steve's authoritative position, which they have watched him grow into and which contrasts with his gentle and unassuming personality.

Teasing stories and stories about teasing are creative and ironic, which is evident when Ted explains how teasing fits into everyday storying on a more mundane level:

> We're talking a lot about just what's going on in our lives, what we did over the weekend, what we're doing, what our kids are doing, what movies we saw, and, oh, whatever degree of teasing or ragging about it [his wife Diane and Charlie laugh] to the person who's talking, to not maliciously belittle them, but if you can get it in there while you can, you know, you [Diane Kroemmelbein: slip a good one in]. It's like a wolf pack: you know, if one limps a little bit, the other ones will get right on 'em [all three laugh]. So, a lot of that, good-natured, never anything nasty. There's no meanness on the island. That stops at the bridge. (June 15, 2013)

Besides underlining the idea that the bridge marks the boundary to a place with kinder behavioral standards, Ted also links the process to the aggressive nature of animals. Ted's tone and his wife's and my responding laughter indicate that not only is the teasing supposed to be entertaining, but he is also joking about the viciousness of the verbal attacks. Sue Meserve calls the larger process "good-natured trash-talking," which can cover a range of teasing about events happening both on and off the island. She recalls,

> Oh, man! It starts, I mean, if Steve took exceptionally long to mend the net, that'll come up. John'll start carrying on about the same thing seventy-five times. You know, how come you didn't, how come this didn't, how come so-and-so, you know.... It's just all that kind of stuff. It's not really a story, it's just, you know... (October 3, 2009)

Sue qualifies this practice as "not really a story" but is reminded of teasing during a conversation on storytelling, perhaps because of its interaction with stories and the creative features it shares with storytelling. The teasing might start with backstory, so that the listeners will understand the reason for the kidding, and anyone who is not hypersensitive and will understand the play can be the subject of kidding. An exaggerated and dramatized tone indicates to listeners that joking is going on. This verbal art is creative in both its verbal reenactment and its stretching of the truth; the people in the community become characters in artful joking, building the "story" in conversation. Indeed, teasing is one of the few occasions when there *is* exaggeration, and thus any criticism the teasing might carry becomes more like play and make-believe, thus preserving good relations.

During an interview with Muriel Meserve and her mother, Muriel told a story about when she and her friend Sis became the butt of the joking. To understand this story, one needs to know that a "live box" is a sort of a cage built of wood that floats in the creek, mostly submerged. Fish caught can be held in the live box to swim around until their buyers arrive hours later. Muriel's opening phrase starts the theme of exaggeration and shows that artful kidding has gone on at the fishery for generations:

MURIEL: This dates back, you know, five hundred years. Well, we used to sell fish during the day when nobody was, you know, when the men were gone.
NELL: Eek and Dot used to sell.
MURIEL: They'd sold them at night with you [Nell: Yeah], I mean, you guys had that covered, but during the day [Nell: You did it] when nobody was over there.... And, as I said, Sis lived there, and we were there, and so, we would go over there and do it. Well ... this one time they had a bunch of carp that they wanted to save for this guy, and—of course they wanted them live—and said, put 'em in the live box, and we were supposed to get them out of the live box and sell them to him. Well, we couldn't get those things out to save—I mean, forget this ... dipping them with the [small net], we ended up both of us in the [Charlie: They're huge!], oh, I know, and you couldn't get them with this net they thought we were going to, I mean, it *sounded* so easy to get in there, get them out, wrap them up, you know.... So [laughs] we *couldn't*, so we really, and you know, and not wanting to give up, we both ended up *in* the live box, *with* them, soaking wet, now trying to *catch* these fish. This guy had a really good show [Muriel and Charlie both laugh], and this story got better and better every time these guys would tell it, you know, because we finally had to *tell* them what happened; we never did get all the fish out.
CHARLIE: And so they weren't even there, but they're telling the story! [laughs]
MURIEL: They weren't even there, but the story got better and better with—'cause they could, obviously, in their minds, between the two of us in there trying to [Charlie laughs] get these fish out, it's pretty good. And [laughs] so we said, we are *not* selling anything live ever again. We'll do it under the cabin. [Muriel and Charlie laugh] ... But, yes, so, you all redeemed us [Charlie: We, *we* were redeeming, that's funny!] as far as the fishery goes, 'cause Sis, Sis and I were done. [Muriel and Charlie laugh] We'll be selling *dead* fish from now on. (October 5, 2009)

Muriel's story is artistically told, with a neat introduction and conclusion to frame the story as an episode in the history of individual women working at the fishery. This story about Muriel and her dear friend Sis became a comical microlegend, but it has also been wedded to everyday storying in the form of teasing and exaggeration. In fact, the increasingly exaggerated retelling of the story of Mur and Sis has become *part* of her story. This story also exemplifies reminiscence as a form of everyday storying. At the start, Nell and Muriel recount the women's roles, an activity that relates to the wider topic we were discussing in the interview. This recitation also resembles the reminiscence that goes on among the crew when a retired crew member returns for a visit, and similarly Nell, Muriel, and I represent three generations of females working at the Lewis Fishery.

Reminiscence sometimes gets interspersed with biographical and character anecdotes that build a cast of characters on the island whose presence and stories create a community. Ray Cashman explains that amusing anecdotes about others become stories about the storytellers and audience members themselves, "and the spectrum of ways of being human in the midst of shared conditions" (2008, xx). At the fishery, this is perhaps clearest in the subgenre of character anecdotes about captains, but stories about other crew members and family members come up during everyday storying to flesh out portraits of important figures on the island or memorialize their contributions. For example, if the rock-paved path or areas for pulling in arise in conversation, then someone may recall the event of Uncle Dory building them. Sometimes others in town with a special connection to the island might inspire character anecdotes, such as Ricky Marriott, who was a neighbor of the Lewis home closest to the island, brother of former crew member Jack Marriott, and Vietnam War veteran. Before his death in 2015, Ricky contributed to the haul most nights through small but essential tasks like catching and tying up the boat when it's brought around the point of the island during the haul, or holding the leash of a fishery dog or the hand of a fishery child. The quintessential example of "local color," Ricky enjoyed playing with tourists with his stock phrase, "I'm not from around here," and he tells stories about these interactions later, as well as stories of his experiences in the military—the more strange, absurd, or funny, the better. Because of Ricky's colorful personality and entertaining storytelling ability, some stories about him begin as PENs but then become part of the group's repertoire of stories. As a stand-up comedian at a local bar, he also tried out his repertoire on the crew at times.

Another area where stories generate and then spread is the quirky subgenre of character anecdotes about dogs, which intermingle with "how's your day?" stories when the "you" in "your" is plural and pertains to both the dog and its person.[1] Many locals come to the island with their dogs, where they can take well-behaved dogs off leash and give water dogs access to the river for a game of swim and fetch. Therefore, stories about dogs often originate on the riverbank, where people who don't know each other watch the dogs and laugh at their antics. With so many dog lovers around, including the island family, these conversations quickly turn into story-trading sessions about particular dogs. Pam observes:

> The dogs, that's the other thing that comes up is the dogs, the different dogs. If something that's happening with Phin, [someone] might chime in about Benji and Blackie, and the different dogs that they've had, like Shira, you know. That'll come up at different times. And then ... so many of them are true fishing dogs, and, and how this Lexie is now starting to turn into a fishing dog, and, those things come up. (October 4, 2009)

Just as there is a line of fishing crew members, there is a line of island dogs, starting with Blackie, a German shepherd seen in an oft-displayed series of pictures of the fishery originally published in the *Newark Sun* in 1937. The dogs Pam mentions here all belong to four generations of the extended Lewis family. When asked about the types of stories she hears on the island, Keziah also remarks on the prevalence of dog stories and describes the conversational practice that Pam enacted during her interview: "I think there was this one. It was like Black or something? I don't know what its name was, but they have pictures of him. And, so, if someone starts talking about one dog, and then they'll like go down this line of all the island dogs there were" (October 3, 2009). This litany of dogs will usually mention Blackie; Bill Lewis Sr.'s dog epitomized the "true fishing dog" because he could identify a shad. The shad he would leave alone, but other fish he would try to catch and shake. In addition to not damaging the prized shad, fishing dogs must be able to restrain themselves from eating scales (which could kill the dog), not go running off and distract the crew members during a haul, not get tangled in the net, and be comfortable around people, other dogs, and boats.

Visitor dogs on the island might be distinguished by their penchant for swimming in the river, but discussion of this talent will quickly move to other aspects of a dog's character. Dogs foster everyday storying by the nature of human-human interaction in the presence of dogs (Robins, Sanders, and Cahill 1991; Jackson 2012; Bueker 2013). Sociologists and others studying American dog parks have found that dogs serve as "social catalysts" for developing relationships between humans (Peter Messent's 1985 term, cited in Robins, Sanders, and Cahill 1991, 13). Douglas Robins and colleagues note that people at dog parks commonly address strangers' dogs the first time they meet rather than addressing the stranger him- or herself, interacting first with someone with lower status (a dog) and thereby avoiding social risks (1991, 9).

Thus, the authors argue, the dog functions as a "bridging device," employing Erving Goffman's term, and the dog is not only addressed but also becomes a safe and entertaining subject of talk (1991, 10, 21–22). Patrick Jackson points out that people tend to know dogs' names before the people's names (2012, 259), and Lewis family members may refer to the people as "Blue's mom" or "Jasper's dad" before human names are exchanged and repeated enough to be remembered and reinforced through conversation.

In his dog park study, Jackson also describes the process of "giving voice," in which people articulate what they think a dog is thinking as a way to show others that they are connected with their dogs (2012, 259). However, different iterations of "giving voice" appear on Lewis Island that make a greater connection between the people than between the owner and the dog. Observers other than the owner may create thoughts and speeches for the dog, using the dogs almost as puppets or avatars—except that the animals' expressions and movements inspire the improvised skits rather than simply delivering them. Here the humans co-create and enact narratives, often humorous ones. Pam Baker, one of the most active agents in Lewis Island community building, masterfully employs the voice-giving strategy to build connections among island guests. For example, she might draw out a visitor by asking questions about his or her dog's personality, eliciting information that can then be used to create the dog's speeches. Her sister-in-law, Sue Meserve, says, "Pam'll talk about the dog. She'll say, 'Look at him! Look at him do—What does he think he's doing?! What is he doing in the net!?'" Pam invites others to join in the narrative game by directing peoples' attention, then asking others to join in the creative narrative process by answering her questions. People jump in to create different versions of a conversation between two interacting dogs or a dog's internal monologue. At times, this monologue narrates the dog's exploration of the island and river, such that the dog's imagined experience becomes the humans' collective development of sense of place.

Indeed, everyday storying about dogs furthers the Big Story of civility. Like family members, crew members, and visitors, dogs are expected to behave with a dog-sized degree of civility. Dogs are expected to be dogs, getting dirty and smelly and barking, but those that are aggressive, do not mind their owners, or interfere with the fishing are required to be kept on leashes and in extreme cases will be asked to leave the island. During my two decades of observation, only one dog owner was banned from the island, a decision reflecting caring for the dog. In that case, the dog owner repeatedly left the dog unleashed and alone on the island, despite being asked more than once to supervise his dog at all times. The family made the decision out of concern for the dog as well as others on the island, for the dog became frightened when her owner disappeared, and snapped at customers and other dogs in her panic. This story has been recounted since with great sympathy for the dog and used as an example of how not to behave on the island. Everyday storying about dogs conveys the island values and reflects the values of the town, which annually produces both a dog parade and a dog calendar. In addition, dog stories serve to connect the people on the riverbank, moving visitors from the category of "general public" into an island community relationship. Typically, these dog owners will return the favor of being able to walk their dogs on this private property by following the town's pooper-scooper regulations, watching out for the property, and engaging in the neighborly communication that underpins the island's community culture.

A similar story type that incorporates strangers into the community is a form of ritual reminiscence I call the "touchstone," which I have heard Pam, Muriel, Steve, Sue Meserve, and Fred and Nell Lewis all mention, although they do not use this term. In this important narrative practice, an infrequent visitor will tell a short story that anchors him or her to the place and the activity in past years. Reminiscence about Lewis Island also spreads beyond the fishery, and Keziah says that she can be anywhere in town and when people hear that she is connected with the Lewis Fishery, they "tell stories of when they went to go watch it [the hauling], things like that" (October 3, 2009). Pam Baker describes a similar process that happens on the riverbank, particularly when a former Lambertvillian is back in town for a wedding or funeral:

> On the bank, you just get wrapped up into the river. And wrapped up into the, you know, "When I was a kid I used to come and swim here," and you'd hear *those* stories, which we can all appreciate, because it's not like it was back when we were little, and these people have this great memory of this town when they *did*—that's where they took their swimming lessons and ... [breaks into dramatizing a visitor's voice] "And when I was little we would go down and watch them fish." [Back to her own voice] You know, so you have these people in their sixties or seventies, these older people, coming down—or then, even people in their forties, 'cause it's something that they can still hold onto. So ... you do have more on the bank, because that river just makes you want to, you know, *gab*, just ... [laughs] You know, as I said, there's something about that, that damned shad, it just makes everybody want to—it's *true*! (October 4, 2009)

Pam covers a lot of ground in this description of touchstone stories, connecting specific reminiscences with the ongoing activities. She points to these memories of this place as an anchor that is reforged through telling narratives *on* the riverbank about the place.

Many touchstones are longer reminiscences of locals whom the extended Lewis family knows, but other stories are from people whom the family and crew do not know, and the story functions in part as an introduction. Touchstones from strangers are usually quite brief and at times sound as if the people have rehearsed them in their heads, or told them to others, such as their grandchildren who have come along for the fishery visit. The topics of these stories include interacting with someone in an authoritative position at the fishery (usually Fred Lewis), remembering details from particular hauls or the island at the time when they came to the fishery in the past, recollecting the adults who brought them to the fishery (in the case of the many people who visited first as children), and preparing the shad at home (in the case of customers).

In addition to serving as introductions, touchstone narratives serve other functions, most notably as verbal touchstones, as relates to a couple of different meanings of the word. The first definition of "touchstone" in the *Oxford English Dictionary* (1a) is, a stone used to test gold and silver alloy. Definition 1b, however, reads, "that which serves to test or try the genuineness or value of anything; a test, a criterion," and this figurative meaning appeared in print as early as 1533. For visitors, the story they choose to tell about Lewis Island shows a genuine connection with the place and its history. Here, the stranger is telling the story as a ritual of authenticity to show that he or she is no newbie but rather someone with a legitimate place in Lewis Island tradition. As more time goes by since Fred Lewis's death in 2004, touchstones that attest to direct knowledge of and interaction with Fred Lewis seem to be told with greater pride. Moreover, stories that reference unique or intimate interaction with Fred suggest even greater specialness or importance.

This interpretation of the PEN as a yardstick of connection holds much in common with Kent Ryden's observation that PENs that occur in a place can reflect group attitudes and emotions associated with that place (1993). At first it seems that the stories only reflect the teller, Ryden notes, but then he asserts:

> Looked at more closely, however, personal experience stories reveal place-anchored emotions which are common to the group of local residents, and therefore to the place, as a whole. For one thing, these tales are rarely told in isolation. They are most frequently told in the context of other similar tales, with one teller prompting another to tell a related story. (1993, 86)

This situation very much fits touchstone telling by locals or former locals visiting the fishery. When talking to other visitors, the locals may exchange stories to find common ground and together define the place, Lewis Island. This process, in turn, builds the "invisible landscape," Ryden's term for applying culture and experience to space, thus creating place.

When nonlocals tell touchstones, there is a slightly different social process going on. To be sure, they may also contribute their stories to define the place and assert authentic connection, but then there is a different communal *process* going on between teller and audience. Usually, people with less connection to Lewis Island will seek out those with the strongest claim to the space to tell their touchstone to. They might tell the story to various people they meet during the visit, but they will often go through a few listeners to get to the most connected audience members. For example, in the field, I often hear touchstones because as a female I am approachable, and because I am decked out in hip boots before the haul and can be seen performing crew roles. When the people talk to me, however, they often ask whether I am "part of the family," then get information from me about where the oldest family member is in the crowd and how the captain is placed in the family tree. By the time they leave the island, they have usually made contact with someone else with more prestige in terms of connection to the tradition. When the family members listen to these stories, they may store them away for future sharing of the kind Ryden describes, but the efficacy of the ritual lies more in the way the family members *listen* to the person's experience. Through the audience role, the extended Lewis family affirms and validates others' experience and connection with the island, thus fulfilling the Big Story of the mandate to share the island. When extended Lewis family members refer to their experience as audience members for visitors telling touchstones, it is clear that they take their role as island stewards very seriously; listening to touchstone stories is a key part of their role in narrative stewardship.

The graciousness of listening is highlighted in an occasional occurrence in which a certain touchstone "fails," specifically an old joke about how to cook shad. The joke starts with a rather elaborate description of dressing the shad, which is the part when the joke teller shows his or her ability to paint the picture and manage the audience's expectations by drawing them into the food prep process. The shad is then nailed to a plank, propped up in front of a fire, and cooked for some number of hours. Again, the joke teller may show storytelling skill by describing the type and age of the plank's wood, further preparation such as nailing on bacon or herbs, and details regarding the fire. After a great length of time, the punch line comes: remove the fish from the board, then eat the plank and throw the shad away. The joke hinges on the shad's reputation for not being tasty, but the touchstone does not "fail" because it insults the people who devote time and resources to shad fishing. The problem isn't even the joke teller's lack of awareness that this is a widely known joke about shad and therefore does not represent a personal connection between the teller and the island. Rather, the teller fails to connect with the island because there is not much of a shad-planking tradition among locals. During the Shad Festival, sometimes a historical reenactor in colonial garb demonstrates the method, but locally shad and shad roe are more likely to be grilled, fried, baked, or—in Asian homes—steamed in banana leaves or curried. Planking is a well-known tradition on the James River in Virginia bound up in an annual political event (Vozzella 2016; Ottenhoff 2017). Shad planking was also done for a Bethlehem, Pennsylvania, shad festival about forty miles north of Lambertville on the Lehigh River, and it continues to be practiced on the Connecticut River in Essex, Connecticut. So, rather than connecting with the place by telling the shad-planking story, visitors who tell that story without any ironic frame instead show that they are disconnected from the island and its shad-fishing tradition. To borrow a phrase from folklorist Ray Cashman, this is an example of "flat textual treatment meets round reception" (2008, 206).

The shad-planking story contrasts somewhat with Mary Hufford's analysis (1992, 92ff.) of a well-known fox-hunting joke in which the dogs make so much noise that the hunters cannot tell what's going on with the fox. The joke is that it's really *through* the dogs' barking that the hunters can tell where the fox is, what it's doing, and whether it's time to call the dogs off to spare the tired fox. In Hufford's case study, the hunters will wait for an uninitiated audience and "perform" the joke together for that audience, whereas at the fishery, the uninitiated audience tells the joke to an audience that knows it well already and knows it does not fit the place. What's interesting is how the fishery family and crew react to the inappropriate story. They usually smile politely or sometimes groan at the "oldie-but-goodie," then divert the conversation to more common everyday storying practices that will better serve to connect the visitor with the place. For example, they might start a conversation that elicits personal experience narratives about experience with shad or the river. Covering new ground, a question about where the visitor is from might evolve into something like "Jewish Geography" (Joselit 2013), a traditional Jewish game in which players find out how they are connected socially through the people, places, and groups they know in common. Finding these connections frequently involves sharing more stories, which become shared narrative experience.

Sue Meserve comments how much she appreciates hearing the customers' "neat stories" (October 3, 2009) on the bank during the haul, or under the cabin during the selling and data collection processes. In these exchanges, we see a place-based and socially based version of what Ray Cashman sees as a main function of the character anecdotes and other stories he studied in Northern Ireland: "Given that narratives are often commemorative orderings of previous happenings, everywhere people tell stories to depict a meaningful past they can use to assess their present and to bolster themselves as they meet an uncertain future" (2008, 1). Visitors negotiate their present placement on the island and use stories to meet the socially precarious position of showing up on private land unannounced. Meanwhile, family and crew use everyday storying along with other story types to reassure the visitors they're welcome, as well as to meet the fishery's uncertain future in terms of ecology and history. Together, both groups engage in a system of richly layered and interwoven narrative practices that characterizes island conversation.

SIX

Talking the Walk: Processional Storytelling

The phrase "walk the talk" became popular in the late twentieth century as a version of "put your money where your mouth is," another cliché meaning to have one's actions support one's spoken values in an enactment, not just a show, of commitment. Throughout this study, we have seen how people involved in the Lewis Fishery express and live their values. In this chapter, the focus becomes more practical, as careful analysis and step-by-step interpretation connect the haul seine process with narrative practice and physical space. Speech becomes action in the concept of narrative stewardship performed during hauls: crew members talk (during) the walk, for a fishing haul would seem incomplete without verbal interaction. Storytelling does not just reflect and pass values; it *enacts* values and, through what I call "processional storytelling," creates more stories. In short, processional storytelling refers to a process that plays out during the traditional fishing haul by which stories are integrated with each other, with the physical movement of crew and onlookers, and with the place itself. This integration has two basic forms: (1) stories being told at different spots and times among different groupings of people, and (2) people collaboratively *building* stories of the haul, the season, and the tradition *through* their conversations during the haul. The multigenre narrative system at the Lewis Fishery joins story types and processes in a ritualistic social activity that interfuses story, people, place, and time to share knowledge and build relationships.

In this context of casually interwoven narrative types, it is useful to decenter the idea of storytellers and audiences as distinct bodies in a performance frame and embrace instead the idea of storytelling as a social activity. Similar to the way Julie Cruikshank writes in *The Social Life of Stories: Narrative and Knowledge in the Yukon Territory* about oral tradition, everyday storying and telling Big Stories are "better understood as a social activity than as a reified text" (1998, xv), and even the most discrete texts on Lewis Island (anecdotes and microlegends) have ragged edges not tolerable in staged storytelling. With so many different tellers and audiences engaged in switching roles to keep the connections and conversations going, the topics and types of stories frequently shift, a process folklorists Ray Cashman (2008) and Mary Hufford (1992, 146–51) describe in relation to conversations at Northern Irish wakes and among South Jersey foxhunters, respectively. Cashman observes: "Storytelling has long been understood as more than reporting of set types of narrative that one might label folklore. It is a state achieved in the progress of sociability, a behavioral mode in face-to-face interaction" (2008, 69). Hufford describes a similar process among New Jersey foxhunters, who use "introductions," "segues," and "conclusions" as well as "frames," "orientations," and "evaluations" to move between chat and story and keep the social process moving with the narrative process (1992, 146–51). The cases that Cruikshank, Cashman, and Hufford describe and the case of the Lewis Fishery all emphasize a story not as sole product but as part of a larger social process.

In contrast to the way men in Richard Bauman's La Have Island study (1972) compete for attention based on storytelling skill and establish individual identities, the identity being established on Lewis Island is as much or more collective as it is individual, and so the storytelling aesthetics are not characterized by competition so much as by inclusion. Narrative devices further highlight storytelling as connection, such as with the phrase "like I said to so-and-so," which invites the audience into the larger narrative community. Although there may occasionally be a spate of stories that escalate with each addition, the tone is more along the lines of "and-here's-another-related-story-you-can-add-to-your-repertoire" than "I-bet-you-can't-top-this." Crew member Keziah describes turn-taking this way: "Sometimes I'll just jump into other ones [stories] if I, like, have like an experience that relates to it" (October 3, 2009). In contrast with telling yarns on La Have Island (Bauman 1972, 337), Lewis Island listeners may enjoy a particularly good telling, but tellers need not establish verbal skill or story uniqueness in order to get heard; they just need to make an effort to connect. Even first-time visitors are encouraged to exchange stories with other visitors, Lewis family members, or the crew, enacting the Big Story of civility.

At times stories are told almost in concert, a practice that stood out when I transcribed interviews with Nell Lewis, her daughter Muriel Meserve, and granddaughter Pam Baker. These three women are particularly active in using nonverbal attends, such as "m-hm," as well as short verbal utterances (e.g., "yeah," "right," and, a phrase characteristic of Nell Lewis and Muriel Meserve: "that's just it") to show that they are keeping up with the story, a practice consistent with sociolinguist Deborah Tannen's (1990) findings on communication and gender. More than that, though, the phrases and nonverbal insertions encourage the speaker in a way similar to "amens" and other spontaneous utterances from the congregation that affirm the words of preachers and testifiers in African American worship traditions. In addition, the Lewis family women frequently anticipate ends to phrases, repeat a phrase a speaker has just said to affirm it, or make contributions (e.g., examples, qualifications) while the main speaker tells a story, which the main speaker then affirms. Here the storytelling voice becomes collective, as the crew and family identities are. While the practice undoubtedly results from knowing each other's speech patterns so well, it may also relate in part to this ideal of working together, which then gets aestheticized and ritualized through talking in concert. The idea of creating stories collectively, collaboratively, and ritually is central to the concept of "processional storytelling," whereby the process of telling a variety of story types is embedded in another process, the fishing haul, which is ritual-like.

Earlier cultural researchers' treatment of the relationship between stories (myths, in particular) and rituals describes somewhat discrete cultural texts that influence and complement each another, often in a causal or almost mechanical relationship. Writing in the early twentieth century, Bronislaw Malinowski, for example, believed that myth tellers use myths as "commentary" on and "preparation" for ritual; because rituals might arise from a "mythic event," they "justify" that event (Bascom 1983, 165). At times, the myth-ritualists' examinations of cultures far removed from their own (including cultures that existed centuries in the past but no longer) seem to arise more from hypothesizing than observation. In comparison to the myth-ritual pattern, on Lewis Island storytelling intertwines narrative and action more tightly and immediately in terms of practice, but the meanings of texts and movements may be connected more loosely. Narratives about past happenings told *during* the haul might not interpret the current haul, and these stories are interspersed with narratives participants develop *through the process of* the haul. Thus, processional storytelling combines narrative with patterned physical movement and thereby combines enactment, ritual, and meaning.

The haul seine method is spatially and temporally anchored in prescribed movements that bring roughly the same people or groupings of people together in the same spots at regular intervals. To see how this works, it helps to break down the haul into its parts, linking each to the story types and storytellers typically involved at each stage. Let us apply the theater concept of the "French scene," which refers to a division in a play's action marked off any time one or more characters enter or leave the stage. That is, if a play begins with three characters on stage and one leaves, we now are into the second French scene, and if two more characters enter together, a third French scene has started. Each combination of characters together in one place constitutes a French scene. The term helps break down one-act plays for rehearsal schedules or classroom discussions, but it also reveals the social nature of the play's action: what happens, happens and means what it means because of who is present, acting, speaking, and hearing. At the Lewis Fishery, storytelling intersperses with decisions and instructions regarding the work. For example, one crew member might ask another, "Should I take the life jackets down, or did Adelaide already do that?" Or one might ask, "Is Steve planning to bring in the boat by the house or down by the cabin?" The answers and details about the day's fishing will then spread through the crew, mixed with short narratives. During a fishing haul, the nature of the work, the business, and the tradition of visiting on the riverbank determine the French scenes in a series of separations and reunions, with patterns of storytelling in each scene—which themselves can occur simultaneously in small, distinct places within the larger space.

A morning's or evening's haul starts as crew members, visitors, customers, and family cross from the parking lot on Lambertville's mainland over the footbridge and onto Lewis Island. At any moment and in any spot, greetings may erupt into "how's your day" conversations, with the more involved conversations among crew members up in the crew's cabin while they trade street shoes for hip boots, perhaps switch into fishy jackets and hats, and add gloves. If a crew member arrives who was not present the day before, one or more people will recount the previous day's happenings, stitching that crew member into the collective experience of the team. After crew members finish changing, they walk down to the island point to assist with mending the net and loading it onto the back of the fishing boat. Short conversations may happen during the stroll between the cabin and the point—for example, acknowledging visitors and customers present—but stories become more involved and offer greater variety as the crew settles into the work of paying the net onto the boat and looking for holes, then pausing for mending. Meanwhile, customers make contact with the woman with the clipboard to put in orders, and other visitors approach with crew members, family members, and each other for "catching-up" or "touchstone" storytelling.

At this point, any reporters or photographers, particularly during the week before the annual Shad Fest, make their presence known, talking to the captain to get an idea of the haul's process and calculating the best vantage points for photos. They may do spot interviews with the captain or other crew members, sorting the crew into their roles and relationships, and after taking photos recording the names of the "characters" in their own developing narratives in the form of articles or online photo galleries. They solicit the Big Stories as well as recent happenings that will give their news stories meaning and the edge of immediacy. Captain Steve Meserve points out that stories are traded both ways on the island. The stories themselves become valued products that people offer but also ask for, and this last pattern obviously appears with journalists. However, Steve points also to a narrative economy that does not involve professionals. He says of other visitors,

> I think people that are not part of the crew also see it as an opportunity to get stories, or to ask the questions that lead to stories because there is a lull—and a lot of times it's just the act of what we're *doing* will start the questioning, you know: What are you doing? Why are you doing it? If it's somebody who's very new to it, they've never seen anyone mend a net before, you know, all they watch is *Deadliest Catch* on Discovery, if they do any watching of fishing. And then they see us doing it, and it's a completely different thing. And the people that'll get real into it—and most times it's people that are even interested enough to come down that'll be interested enough to hear the history and the story of it. (October 3, 2009)

As one would expect, these lengthier conversations tend to take place during the least active points in the fishing, starting during the gathering, mending, and loading. The storytelling then continues with visitors, but may shift away from the most active crew members during the transitions in the haul when crew and family members part and rejoin in various configurations.

After the crew loads the net, they pull the boat around to the side of the island and often pause a few minutes before the first separation, when the one crew member, usually a woman often accompanied by a child or a dog, gets into the boat to steer. Other crew members then grab hold of the rope used to pull the boat upriver, and the ritual nature of this walk upstream becomes visible when some members drop off the rope so as not to shorten the lead and ground the boat on the bank. That is, some crew members take part in the walk upstream, even if they're not needed to pull the boat upstream, or at the upstream location for the next stage. Meanwhile, some members of the family or crew—usually women and children—stay near the island point to interact with customers and visitors and rake the spot where the fish will be hauled in.

During the first separation, the visitors and customers begin to chat on the bank with each other and the fishery women who have "stayed down below," and continue to do so until the fish are hauled in. Pam Meserve Baker's fishing memories center on these conversations on the bank, for as a girl she was not encouraged to fish as the girls are today, and so she played more of a "hostess" role for the visitors on the island while the crew was upriver, and thus made connections with older generations:

> It's nice, now, even though I don't have . . . the memories of the actual fishing, but I've got millions of memories of my grandfather, and all the guys who used to be on the crew, and these old timers come like Mr. Hentschel when he used to come, and just talking with him on the bank . . . and just listening to them, and this fish that [smiling], for whatever reason . . . it's a combiner of people; you know, everybody likes to talk about the American shad. (October 4, 2009)

Her memories reinforce her brother Steve's message that story trading goes in two directions. The narrative discourse doesn't just run between crew and family on the one side and visitors and customers on the other, but also moves from visitor to visitor, visitor to family member, and so on. The Lewis Fishery factors heavily in the storytelling, but the Lewis family and crew are not the only experts, nor always at center.

The people in the boat and those with the rope moving up the path sometimes call out to each other as the crew pulls the boat upriver, but usually wind and water scatter the sound too much for meaningful conversation. Instead, the two groups carry on parallel conversations about their work days and creative work projects, and about the various signs of spring they see as they move upriver: particular spring flowers in bloom, particular broods of ducks or geese, the height of the poison ivy. Even *within* the line of crew on the rope, though, communication is a "challenge," as Dan Tuft says:

> People will be yelling something up either ahead of the line or back in the line, which always makes it a challenge, because if the person in front of you is talking to someone behind you, you can't really tell what they're saying. It's hard to hear them. So usually just the two people who are closest to each other would say something to each other. (October 3, 2009)

The small cluster of an adult and one or two children makes the boat a better site for conversation. Sue Meserve remembers when crew members Sarah, Keziah, and Adelaide were small and went up in the boat so that they could be supervised and not slow down the crew:

> For years it was Keziah going up in the boat with me, and Keziah and me talking, either talking about fishing, or talking about people, or talking about school. And then it got to be Keziah and Adelaide together. And for a while Adelaide went through a phase of, you know, tell me about when you were little, and what did your brother do and what did your sister do and tell me stories about that and all. And now it's gotten to be Miss Social over here [jokingly refers to nine-year-old

Adelaide doing homework in the next room during the interview] is telling me what went on with her day and who did this and who did that. (October 3, 2009)

Keziah also remembers regular stories about projects Sue was directing at the theater shop she ran (October 3, 2009). When Keziah was in fourth grade and thinking of an engineering career, in the boat Sue would tell stories from her shop in installments, which she framed as engineering puzzles, to be continued the next night on the boat. Sue also recalls looking for signs of spring with the girls: "We look for the daffodils on the side. We look for ducks; we look for the little baby ducks and the baby geese.... We've been seeing turtles every once in a while now. On the walk back, we're looking for Dutchman's breeches, and we're looking for Jack-in-the-pulpits, and May apples come up" (October 3, 2009). Like anywhere else, the stories relate to the people sharing them, and in these cases, they cater to the interests of the children, who don't have to compete with other adults for attention when they have Sue alone in the boat.

The crew then reunites at the furthest northern point of the haul. The women and children jump off the boat, while the men assigned to rowing and poling the boat, including the captain, get into the boat. Often this is a time for joking about who will do the most physically challenging roles, and some stories that were in progress along the path may continue until finished before the boat sets out. This stage in the haul may grow or shrink in length, and the storytelling with it. So intertwined are the social and natural processes that a community obligation after fishing (e.g., voting in the school board election) or the weather (harsh wind or rain, threat of lightning) can influence the amount of time the crew jokes, completes conversations, or trades information about signs of spring.

At the next separation, the crew splits into three or four parts for simultaneous French scenes. Usually it's men in the boat as well as filling the landsman role, dragging one end of the net back down the outside path. If there's time, women, children, and a dog might head down the center path instead of the outside path. A couple additional male crew members head back down the island either on the outside path or on the center path. Again, the weather influences the time the group converses separately, but so too will the river itself influence the narrative activity. When I mentioned this idea to Steve Meserve in the fall of 2014, he immediately thought of the fact that the appearance of the river might remind someone of a particular story, which they will then tell. However, the river governs narrative behavior in more profound ways than merely triggering memories. A high river moves quickly, so the haul will need just enough net for the crew to get out as far as possible and still bring the net in quickly enough that it does not just spill onto the shore into a tangled mess. Similarly, if the river is very low, the crew pulls in earlier, on the side of the island rather than farther down on the point. In both cases, if the crew does not rush down the bank to meet the boat and grab the sea end of the net to start pulling it in, the current will push the net onto the bank in a big, tangled mess.

If the river height and speed are somewhere in the middle, crew members in the boat and on the island can use more net to get farther into the river channel and take more time bringing in the net. Typically, those going down the outside bank walk down and sit on a park bench to talk to customers, visitors, and other crew members while they wait for the boat. The landsman goes down the bank at about the pace of the river, always arriving at the point last, when much of the net has already been pulled in. Usually he (and it's usually a man) starts alone or with one other crew member there for backup if the bank is slippery, but the captain meets him close to the point to finish the haul. Occasionally, however, a very curious visitor follows the crew up the bank, then comes down with the landsman, who will likely share the Big Stories and a microlegend or two. Meanwhile, those on the inside path may take more time moving down the island, pausing to take a look at Mrs. Berg's garden or blooming wild flowers, the adults periodically looking out through the trees to the river to gauge the progress of the boat.

Curiously, the two small crew groups at the furthest eastern and western points—the men in the boat and the women and children on the inside path—tend to have parallel conversations. Dan Tuft describes the boat conversations this way:

> Sometimes those are the more private conversations or private stories about something that happened during somebody's day, or trouble that somebody's having, or family things that are going on. I've noticed that those tend to be more personal stories out on the boat. (October 3, 2009)

Dan explains that while conversation may depend on who rows, the spatial atmosphere also influences the conversation: "It's more private out there, even though you're out in the wide open, but there's fewer ears" (October 3, 2009). Steve and Sue Meserve describe something similar in their interview, referring to Steve's nephew, Andrew Baker:

> STEVE: Andrew has told me some things out in the middle of the river that [laughs]—

SUE: Yeah [laughs], true confessions by Andrew out in the middle of the river...

STEVE: Yeah, yeah, it makes for interesting hauls. Sometimes the conversations can get too engrossing and we all of a sudden are a little bit too low [too far downriver], and should be turning in already, or something like that, so we have to watch that, so I've always got to keep an eye out to exactly where we are in the river, so we don't get into trouble. (October 3, 2009)

In both cases, the remoteness of being in the middle of the trees or the middle of the water encourages talk of a more private nature, often troubles-talk in both spaces, although the people get divided along gender lines, females in the privacy of the trees and males in the privacy of the boat. As Steve points out, the river and the process still control the storytelling, because once the women come close to the fishing cabin they are again in a more public space, and once the boat approaches land, the work gets too hard to talk through and the pace switches into high gear for a few minutes.

When the boat hits land there is a flurry of position- and life vest–switching while the net and rowers leave the boat, the crew quickly pulling the sea end of the rope up the bank to keep the net close to land, and others—including a terrier—boarding the boat to row it around the point of the island to be tied up. Once the boat lands back on shore and all the crew members except the landsman are present, the conversation settles into a groove again, sometimes with joking among the crew that entertains the visitors, who by this time have gathered into a recognizable audience.

Interspersed with the joking emerges an important part of processional storytelling: the collaborative creation of stories. At this point, those who have been separated ask about and share observations, as if they had gone on separate voyages. They say: "Did you see that mama duck and her babies? There were six yesterday and I could only see five today; did you count six?" Or, "The May apples are just starting to open up now." Here, the crew develops three different stories: the story of the haul, the story of the unfolding spring season, and the story of the fishing season. Ted Kroemmelbein and his adult stepdaughter, Kelly McMichael, recall seasons upon seasons of tracking signs of spring, much like Sue recalls doing with her niece and my daughters half a generation later. Ted recalls they always had "cool stuff to look for, because that's when we always notice when the violets are coming up, because they always came out on the island before the yard for some reason. And that we could always see the geese, and the little [Kelly: babies], and the little eggs, and then see the babies" (June 15, 2013). Kelly anchors the season to the island by recalling the precise place to find goose nests: "That was another thing that we always used to do.... [It] was on the one side of the house [that] they used to always make their nests ... you couldn't get too close, obviously.... But we always knew when they were gonna be born" (June 15, 2013). The springtime rebirth story is integral to the fishing, not only because the springtime shad run is part of this process and the fishing itself is a "season" that happens in spring, but also because being out in the spring weather forms part of the crew members' motivation to volunteer their time.

In addition to the story of the spring season, a story of *this particular haul* emerges as crew members share with each other—and at the same time with the audience of visitors—anything distinctive about the haul. Amid observations of signs of spring, they might exchange remarks like these: "It was nice to get out of the wind going down the center path; how bad was the rowing?" Or, "What was happening that it took you guys so long to get in?" "Oh, we got stuck on a log under the water brought downriver by Sunday night's storm, and Steve actually had to pull us backward on the net to get it unhooked. Hope we didn't lose any fish then." By updating each other on what happened when they were apart, the crew actively builds the story's "rising action" (to borrow a literary term), discussing how earlier events might affect the catch. As entertainment, this collaborative storying builds suspense as crew and visitors alike anticipate the final results of the haul, the "climax" of the story of this haul in particular.

When the landsman rejoins the group and the final phase of bagging up slips into gear, story*telling* pretty much stops, but story-*building* continues as everyone tries to predict what will turn up, calling out what they feel pulling on the net or what they see, such as a fish jumping over the net or water roiling. In the last few yards of pulling in the net, splashing increases, and then often cheering erupts from the crowd and crew after the last pull of the net brings the fish onto the cement. At this point, talking stays focused on the task of deciding which fish to keep and which to throw back, then counting those fish thrown back. Instrumental speech (speech to accomplish some purpose) dominates, and narrative is pushed aside in the hurry to safely get the unneeded fish back into the water. The count then becomes a sort of musical marker in the larger ritual. The captain and two or three other crew members negotiate with the head fishwife which fish to put into the buckets for sale, then call out to the girl or woman with the clipboard the names of the species thrown back: quillback! gizzard! gizzard! cat! bass! five gizzard! eight cat! and so on. Those crew members not throwing back fish either hold the net, or back off from the net slightly and draw away the most talkative visitors, for often visitors perceive the female with the clipboard as being unimportant and thus interruptible when she actually plays the most difficult and important role at that moment. Thus, by building the story of the haul with a potentially disruptive visitor, other crew members support the process of the haul.

After emptying the net, the crew splits again, and narration resumes more generally. One part of the crew pulls the net back a few feet above where the captain expects the water height to reach over the next twenty-four hours. On some days they actually pay the net back up onto sawhorses to dry out over the break between hauls, or to keep it safe if weather predictions suggest a possible pause in the week's fishing schedule due to high water. At this point, these people have time to interact more with the visitors, and another round of collaborative storytelling takes place. Captain Steve says that they "do the postmortem on the haul, if it was good or bad. And, you know, relive the highlights" (October 3, 2009), a process that continues through the evening for the crew. Directly after the haul, however, the story of the haul is recounted for visitors, now including the information sharing and narrative building that went on between the various crew members during the early stages of pulling in the net.

At this point, sport fishermen often present themselves to the male crew members as colleagues ready to discuss the fishing, often with that competitive edge to the conversation that sociolinguist Deborah Tannen (1990) sees as part of masculine culture. Crew member Ted Kroemmelbein uses humor to express discomfort with fishermen's competitive speech patterns:

> Well, there's a lot of fishermen who all know everything. Whether they do or not, you know, remains to be seen. It's always debatable. So, ahhh, they end up grilling us about what's going on more than anything else, so I think a lot of them are there for the information. What are you catching? How are you catching? What do you think blahblah-blahblah-blahblah-blahblah-blah? (June 15, 2013)

The tone Ted describes as "grilling" in this kind of information exchange sounds less amiable than that in conversations with people who are interested in the Big Story of tradition or with folks primarily interested in springtime along their river. While Ted will always answer the visitors' questions politely, he clearly does not enjoy competitive conversations with fishermen as much as other conversations.

Nevertheless, whether the conversation more resembles teamwork or a contest, the crew members and various visitors to the island collectively build the story of the season—but this time they discuss the fishing season more than the natural season. Visitors might ask about the season when they first arrive on the island during the haul, but after the haul, visitors who arrived when the boat was out ask whether it's been a good season. Visitors join with crew members to build this story of the season, crew members adding the new haul to the others that came before in this and other seasons, while some visitors chime in with what they've read in the newspapers or their own experiences pole fishing in other spots.

While the talking and net handling commence, the baskets of fish are hauled away to sell under the cabin, followed by a few customers and other visitors. Sue Meserve says, "Yeah, we're very anticlimactic. They don't understand; I mean, the fish come in, that's it, you know. And there's a special group that knows to come up to the cabin and catch the late show" (October 3, 2009). Other than the customers, the visitors who come under the cabin to see the next steps are folks who have made a special trip to learn about the fishery, are visiting the fishery for the first time, or have brought children along. Most others, including reporters, do not follow to see the next steps because they anticipate and avoid witnessing the fishes' inevitable death, or because they went once and know that the space is small and the work is central, or because they focus on the more masculine activity of fishing. At times tension hangs in the air as the fish are sexed (identified as male bucks or female roes), and if the fish in the net are fewer than the total the customers requested, the fishwives determine who will get how much of what they ordered. There may be a pause to give visiting children a chance to get a close look at the fish and even touch them. After that, the fishwives set up a rhythmic process of taking data and scale samples, wrapping the orders, making change, chatting for a while with customers, and saying goodnight. Again, a sense of musical ritual emerges as those taking data call out the sex, length, and weight of each fish, then pause to transfer the scales into the tiny envelope on which the data has been recorded. The captain and one or another of the crew might pop down under the cabin to learn the final numbers of kept shad sorted by sex, thus reprising the ritual of the count. This last detail not only completes the story of the haul but contributes to the story of the season: the size of the roe fish, the color of the eggs, whether the roe "hold onto their eggs" or the eggs are "spilling out," and the proportion of bucks to roes, which may suggest how far off the end of the spawning season is.

Contrary to the overall Lewis tradition of rewarding good behavior, the occasional impatient or unpleasant customer will likely be handled first, so that the other customers, visitors, and crew members can relax and enjoy a good chat. Just as a low river height will mean a more leisurely pace and more storytelling, a low number of customers will also facilitate more storytelling under the cabin. Sue Meserve's estimation of narrative activity at this point offers a different perspective on crew-visitor interaction from that of the people who hang back to talk to the crew putting away the net. She says with great interest,

Oh, there's all kinds of storytelling going on under the cabin, and it depends on how many people and what they want to know. For the most part, the people who are customers are really patient. And a lot of them have been coming for years and years and years, so they actually have their *own* stories to tell. And they'll ask about family: how's Grammy doing? And they're always trying to get who's related to who straightened out, and so that can be a conversation in itself for half the evening. And then there'll be stories about, you know, how do you cook the fish? And how do *you* like to eat it? . . . And, what's the best way to clean it? And what kind of fish is that? And do you, you know, do *I* [meaning Sue herself] eat that? Errr. And then there's a couple of people that'll hang around and who will just share whatever story, like this guy, David Fish. He'll hang around, and he'll talk about bringing somebody over: he's got a brother-in-law or somebody he's going to bring over for fish. Now the, the sisters [two women who are inseparable and look enough alike to be sisters] are starting to stick around and talk a little bit. . . . [They] tell us about family or tell us a little bit about, you know, who they shared the fish with or something. (October 3, 2009)

In Sue's description, we see that the conversations here, whether the customers are male or female, resemble the interactions between visitors and fishery women on the bank more than they resemble the conversations between sport fishermen and male crew members. Even though the conversation might tackle the topic of how the season is going, under the cabin that topic takes a back seat to everyday storying that makes connections, and sharing accounts of personal or cultural traditions.

Once the selling stops, the women and girls head upstairs into the cabin with the men to change back into their street shoes and jackets and hang up their wet gloves. The ritual of rehashing the count repeats in the cabin, as does the story of the haul. In this part of the processional storytelling, the crew rejoins for the last time, and together they fine-tune the story of the day's haul when the women and girls add information about interactions with the customers and other visitors during the selling process. Keziah observes, "At night I think, [the stories are] more interesting; there's less people. It's the people who stick around the longest. . . . They're more interesting stories. They have more substance to them. They're not just like joke ones. You talk more about what's going on [in the crew members' lives or the community] . . . or . . . the fishing, something like that" (October 3, 2009). Storytelling among those who linger may get into in-depth comparisons with years past. Conversely, they will then look forward to the next night's fishing or other activities, and, to quote Steve Meserve, "just general chatter too, just almost like any other workplace, really" (October 3, 2009). As crew members change back into nonfishing shoes like Fred Rogers switching back into his blazer at the end of *Mister Rogers' Neighborhood*, we see another reprise of "how was your day," which incorporates the haul that just happened as part of the day—now a story of the past for the first time. Talk of the next day's plans serve as foreshadowing for the next day's round of everyday storying, before final goodbyes ring out on either side of the footbridge.

On the way back to their cars, crew members recombine as family units, with parents resuming full charge of their children, as opposed to earlier in the evening, when the captain wields authority over the kids' tasks. Like the kids rejoining their parents, Steve and Pam also head across the creek to stop in at their mother's home on the mainland before heading to their own homes. The nightly ritual includes the exchanging of surveillance stories—things Steve noticed that his mother should look out for the next day—as well as the final word on the night's haul. Steve gives the final numbers to his mother, who owns the fishery with her sister; in the past, he gave this final nightly report to his grandmother, and to his grandfather, Captain Fred, before her. Steve describes the ritual of the evening's final report in 2009, when he still gave the report to Nell Lewis, his grandmother:

> We go in, *I* go in, and I let her know—even if my mom, my sister, or John [his brother-in-law] might have gone in first, and ... it's coming from us that the official numbers are, and she needs to hear it. Now whether or not she really needs to hear it or not, that I don't know. I think she does. I think she likes it. I think she enj- ... she needs to, to know that she's still considered a part of it all. And ... you know, it's paying respect to her and her work over the years with it, because she spent as much time over there as my grandfather did, keeping the ledgers of fish sold and who was ordering. (October 3, 2009)

Here, Nell plays both the role of owner, which she inherited after her husband's death, but also continues her fish sales role, which she had all but completely passed to Steve's wife, Sue. Sue adds to Steve's description of Nell's response to the daily report, which emphasizes customers:

> Yeah, she wants to know, did you find Mrs. Cooter? Did this one show? Did that one show?— 'Cause they called during the day. And, you know, y-yeah, you have to answer to all of that, and tell her that we caught 330 catfish and no shad, you know. Have you seen so-and-so this year? You know ... (October 3, 2009)

Muriel Meserve gives a similar recounting of the final report but mentions the surveillance stories, because when Steve goes home for the night, she watches over the island. What's more, when I asked Muriel for stories about "surveillance," she combined the safety surveillance stories with social ones:

> The surveillance stories—there's actually a two-pronged thing there as far as people are concerned. The surveillance thing, um, now that nobody's [living on the island], Steve [monitors and reports on it] more often, just to check to make sure nothing funny's going on and that we don't need to call the police to come check it out now and then. But there's another kind of story he comes back and tells too, as far as who's been there that night to see the haul, because there are these people who have come for years and years and years and years, and, so, he'll say, you know, Mrs. Sanders was there tonight, or, . . . you know, who's been there, and what they had to say, which Dutta families came, because there's the Duttas and the Dattas, and a lot of them will call and so he'll tell me which one it really was, and who was there, and what they got, but . . . there's this whole . . . group of people that just have been coming for years and years and years; you get the report on *that*, too. (October 5, 2009)

Muriel depends on Steve's report on face-to-face interactions to clarify phone messages left by customers during the day and to complete the story of the haul through the lens of interpersonal connections. Sue's and Muriel's description of the final report reflects the last step in processional storytelling, which holds much in common with other stages: crew and family who were separated recombine, update one another, and together build the stories of the haul and the season, which will be continued again the following day, when the "changing of the guard" will transfer primary responsibility for the island back and forth between Steve and Muriel yet again.

The ritual nature of this final report perhaps appears clearest when Steve notes that others might have already given Fred, Nell, and/or Muriel the same information about the haul, but they need to hear it from Steve. This seems most striking when the earlier speaker is Keziah stopping in at the house before all the packing up is done. Although Keziah is the one who actually took the count on the riverbank and who was there while the kept shad were sexed and counted, and might even have reported to Steve the full number of kept shad as well as giving him the number of fish thrown back, Steve's report based on Keziah's numbers are more "official" than Keziah's numbers for the ritualized final narrative of the evening—as befits a fishery anchored in a family tradition.

Again, here we gain a sense of ritual from story repetition, narrative fragments, and other types of conversation. In some ways, the various bits of conversation needed to execute the haul ("Do you have the vests down there?"

"We're coming in at the steps!") intersperse much like what happens at the Irish wakes Ray Cashman (2008) observes: condolences and stories intermingle, connecting the event at hand with the larger social fabric within the context of community history. At the same time, the cumulative creation of fishery history with each haul resembles the recursive and integrative effect Mary Hufford sees in the Pinelands among foxhunters. There, stories of chases become the context for other chases (Hufford 1992, 11). Both the foxhunters and the fishery crew members use stories from the past to make decisions that affect the chase or haul they are in the midst of, thus creating happenings from stories while creating still more stories from happenings, factoring both back into the sense of place and the ritualistic activity of hunting or fishing.

Having traced how narrative cogs interact *during* the haul to build the narrative *system*, we see how crew members develop certain stories *through* the haul, *through* the process of parting, observing, rejoining, and narrating. People may co-narrate a particular story through turn-taking, but they also co-create larger stories in an emergent fashion. That is, observations (e.g., of net behavior or spring blooms) made along the route of the haul by different subgroups are reported to and discussed with one another at points of reunion. Not only does this create a story of *this particular haul*, but it joins stories of previous hauls and previous seasons to create a story of *this particular season*, both in the sense of spring season within the context of the year's rhythms and also in the sense of this shad fishing season, within the larger story of the species' health and history and the larger story of communal fishing and island traditions.

In a process paralleling the one Hufford sees in Jersey fox hunting, the combination of the haul and the narratives surrounding it helps participants create structures that in turn help them ride out the anti-structure inherent in traditional haul seine fishing on Lewis Island.[1] The wily fox and the quirky shad practically beg for comparison as trickster figures who lead hunters and fishers into foolish undertakings. Pam Baker has a sense of camaraderie with the sport fishermen also outwitted by the shad:

> It just is so cool, because even the rod and reel guys will talk about it, how, [in a gritty male voice] "Oh, I had this great one, and we got seven because they're really bitin'" and they all want to talk about capturing this fish, and even though we're doing it with nets, it's the same kind of thing. They elude us all the time. This past year when it was just all of us old people fishing because Sarah wasn't there, the kids weren't there. Nobody was there and here we are, the five of us or six of us, and it starts raining, and at that time we hadn't gotten any fish I don't think. And I said, "Oh, we're the village idiots, out here in the—," and, and then it

started to pour, and my mother said, "Oh no, we had *years* of that. *We* were the village idiots." (October 4, 2009)

Pam laughs at the obsession with the shad driving her family to spend vast amounts of time in miserable conditions. The shad appear in daily conversations as an ominous "they," as the crew wonders "what they're doing out there" in the net—or not in the net, having outwitted the crew yet again. And yet, speakers understand that imagined conspiracies conducted by shad are themselves a collaborative fiction.

The shad's quirkiness perhaps reflects the unpredictability of springtime, itself hanging in the balance of structure and anti-structure,[2] negotiated through narration. We say that March comes in like a lion and goes out like a lamb, but then April has its lion-like days, and in late May or June a cold front can come in quickly and blow up a storm whose lightning or hail curtails fishing. Ted Kroemmelbein's explanation of what he likes about the island is anchored in his experience of the spring season there:

> It's the same but it always changes. Same thing, I like to see the violets come up, the little yellow things, the little white things, the bluebells, the jack-in-the-pulpits and all that, and that's when they all come up, and the leaves come out ... —and sometimes we're there when it's *snowing*, pulling nets, and go to, you know, sweating bullets a month later. You don't know how it's going to be. (June 15, 2013)

The overall emerging narrative of the season emphasizes a structure that moves from turbulent and cold to calm and warm, but the day-to-day experience of the first half of the season means constantly discovering that one has dressed too light or too heavy and in midhaul there's only so much one can do about losing or gaining clothing. Likewise, the crew sometimes does not know until the last minute, perhaps even after loading the net onto the boat, whether the threat of an electrical storm will be enough to cancel a night's fishing.

Rather than the fickle weather and shad, though, the "real" anti-structural force is the river itself. However, the river is not just an amusing trickster figure; its ability to destroy is a fact that never strays far from the Lewis family's and the crew's consciousness. This is why crew members' children wear life vests during the busyness of the haul even when they are not on the boat, and why they are instructed—often with a recounted microlegend—never to run after a ball that has landed in the water during play. It's why the women who row the boat around to be tied up pull extra hard as they round the island point so as not to get stuck in the eddy. It's also why stones may not be thrown into the river, for they are needed to prevent erosion, hold down the island soil, and protect the island from the constant pull of the river. The reminders of flood destruction stories appear in the worn places on the buildings that the crew walks past during the haul. More immediately and less dangerously, the river's fluctuations cause the crew to constantly adapt their storytelling to the water's ever-changing height. The river's height decides whether the crew must add or remove net before the haul, whether they must run down the island paths or will have time to sit on a bench waiting for the boat to land. In turn, these impacts will determine how many narratives people will tell, how long those narratives will be, how much information a teller can impart to a listener, and perhaps whether or not tellers and listeners will form tighter bonds on that particular evening.

Nevertheless, against this fluctuating background, the nature of the narrative *process* imbued with ritual adds structure to the shifting forces of the river and the passage of time over generations, helping people negotiate the island as a liminal space between river and mainland, and the present as a liminal point between past and future.[3] The structure of the narrative process combines and recombines the various story types, told within and through the structure of the haul. Of the story types discussed here, the microlegends provide the least stability in that they frequently center on some destabilizing force, like the floods or threatening Jet Skiers. Yet even then, the overall story is usually one in which the threatening force is subdued, and even when the island caretakers take a beating, their telling of the story evidences their endurance in the face of adversity. This endurance forms the heart of the Big Stories of tradition and the environment, offering tenacity and civility as stabilizing social forces. And lastly, the everyday storying provides the glue that holds the system together, day in and day out, as the enactment of civility in the form of friendliness and hospitality. The processional stories of the haul and the natural and fishing seasons underline repetition and stability, with tradition creating meaning with each new iteration of haul seining or springtime, and bringing hope that the faithful monitoring of the fish and the river through the fishing tradition will sustain all three.

The stabilizing system of narrative stewardship works through the ritualistic haul, which has common ground with other mobile meditations, such as walking the Stations of the Cross, walking a labyrinth, or walking on pilgrimages. Writing of Glastonbury, folklore and religious studies scholar Marion Bowman notes how Christian and pagan groups compete for influence over multiple religious traditions' sacred space, "using the act of processing, and an array of traditional accoutrements such as statues, banners, costume and song, as vehicles for staking a claim in contemporary Glastonbury" (2004, 283). While less aggressive in their claim to "ownership," visitors to the Lewis

Fishery who set foot on Lewis Island and tell verbal touchstones also assert a connection between themselves, the place, and its multiple traditional stories. Folklorist Torunn Selberg (2006) similarly describes a practice on the Norwegian island of Selja, another medieval Christian pilgrimage site, where event planners use procession to connect people with sacred and historical space. All three sites bear aspects of pilgrimage, and on both Selja and Lewis Island, stories are told or built while walking. However, on Lewis Island, the daily haul creates less of a spectacle than the Shad Fest, and less of a spectacle than events on Selja or in Glastonbury.

Rather, the processional storytelling on Lewis Island better resembles a British tradition with more practical import described by Kent Ryden in *Mapping the Invisible Landscape: Folklore, Writing, and Sense of Place* (1993). Ryden describes English parish priests' practice of walking the borders of a town with their congregation to ask blessing for the agricultural year ahead, examine the properties, and pass knowledge to the next generations of what constitutes the border (1993, 26). While not asking for an agricultural blessing, the fishing crew does seek—and finds—food through the haul. The possibility of a religious interpretation is ever present not only through the Christian fish symbol but also in a customary Lewis family response when asked what they will catch that day: "Whatever God puts in the net." Like the parish priests, the crew examines the property during the walk both for spring growth and for blooming, but also as a surveillance activity to support the family's role as stewards of the island. Lastly, traditional knowledge is imparted to the next generations, both to young adults and children directly connected to the family and town but also to new generations of visitors from farther away. Rather than looking at borders that divide private property like the Lewis family and crew do, however, these generations of visitors are oriented more toward common values and heritage that connect people. This last idea, sharing cultural heritage, is articulated in Keziah's thumbnail functional analysis of storytelling at the fishery: "I think, a lot like of the conversation or stories are all like to, make fun, be funny and stuff, but mmm, I think people tell stories about their days too sometimes to just get stuff off their chest. I think that's it, and also I think a lot of the older stories from the past are told to keep the ... uhh ... tradition going? That kind of thing" (October 3, 2009). Connecting ridiculous joking to sublime self-reflection, this young crew member clearly understands the stabilizing function of processional storytelling and the whole storytelling system.

Taken cumulatively through processional storytelling, narrative stewardship on Lewis Island constantly develops and deepens the storytelling system, making sense of one's place in tradition, sense of the unreliable natural world, and meaning in the community.[4] The system of narrative stewardship on Lewis Island may have fishery family and crew members at the center, but they must also include visitors, inviting them to tell their verbal touchstones, to share in "how's your day" storying, to connect with the larger stories of spring, tradition, and the environment, in order to build a civil community. In her study of flood narratives, Barbara Johnstone notes that "private stories and individual ways of telling stories give rise to shared, public stories, and ... public stories form part of the context from private ones" (1990, 13). While the stories developed collaboratively through processional storytelling are shared through the creative process, all the other stories told on the bank are also shared, with people switching in and out of the teller, character, and audience roles: "It's like I said to so-and-so" mirrors "It's like what so-and-so said to me." While people switch between individual and collective stories, people and place also become integrated with one another and connected to stories through the essential movements of the haul. Yet, although participants build these connections through daily activity during the fishing season, the overarching sense of belonging and community takes time, opportunity, motivation, and commitment to solidify.

SEVEN

Who-All's Coming Down to the Island: Belonging at the Lewis Fishery

One day in the hallway outside my office, I crossed paths with a colleague, Denise, who was practically sputtering with laughter in anticipation of telling me a story. In a casual conversation with her husband, Sam, an avid line fisherman, Denise told him that I was involved with the Lewis Fishery crew and that at work she was hearing about the fishing season. Much to her surprise and amusement, her husband was incredulous and even a little indignant. She told me that he vigorously insisted, "She can't just do that!" He explained to her that you had to be a member of the Lewis family to do it and recounted that the Lewis Fishery was the only operation around with a license to fish using nets. Sam's reaction is not unusual; members of the crew who are not Lewis family members—especially women and children—often face surprise when people familiar with the Lewis Fishery learn we are crew members.

Denise's husband clearly knew the Big Story of the traditional haul seine method and the now unique position of the Lewis Fishery in the region, but she had to recount much more of my story—that I lived in Lambertville, that I had started studying the fishery over a decade ago, and that my husband and children also fished—before he was convinced of the connection. His reaction points to the traditional pattern and assumption in family businesses: you have to belong to the family to participate. While the exclusive "you have to be part of the family" does not work in a family that lives a mandate to share the island, neither is it accurate to say that the doors are flung wide open and anything goes. Again, narrative stewardship plays a role in including more people in the tradition and welcoming them to the place, while instructing visitors in the behavioral expectations that make Lewis Island a civil place that runs on commitment and friendliness.

As Debra Lattanzi Shutika points out, many scholars look at the concept of "belonging" in relation to the lives of immigrants (2011, 94–95), and this becomes an increasingly important issue to consider as a facet of "sense of place," particularly because physically moving from place to place defines migrants' status. It's also important to remember that "belonging" has an important place in social psychology literature in relation to a variety of social situations and identity categories: school (Gillen-O'Neel and Fuligni 2013); work (Cockshaw, Shochet, and Obst 2014); religion (Alper and Olson 2011; Dougherty and Whitehead 2011; Krause and Bastida 2011); socioeconomics (Hoffman 2012; Ostrove, Stewart, and Curtin 2011); gender (Richman, vanDellen, and Wood 2011; Gutierrez et al. 2013; Rosenthal et al. 2013); and ability (Laursen and Yazdgerdi 2012; Hall 2010). Many studies seek to identify barriers and factors that promote the likelihood of belonging. In most studies, belonging and the related concept of "connectedness" correlate with mental health and well-being, some studies beginning with that assumption, based on other studies. The idea that feeling as if one belongs is beneficial has become common sense. Expressions such as "After that, I felt I belonged" or "They acted like I didn't belong" frequently figure in informal conversation about whether one had a positive experience in a social situation, with belonging implying a positive experience.

This chapter examines the various ways in which people belong at the Lewis Fishery; the various degrees to which people belong; what determines whether, how, and to what degree one belongs; and the place of narrative stewardship in this dynamic. Lattanzi Shutika's explanation of what creates a sense of belonging offers a good place to start this inquiry:

> The sense of belonging is constituted through shared meanings and sense of social alliance between people and the places where they reside: it does not necessarily reference a geographic location but can include places that are physical, virtual, or imagined. (2011, 15)

Anchored to a specific place by a fishing license and tradition, the Lewis Fishery primarily links sense of belonging to a "geographic location," and chapter 8 will fully explore sense-of-place issues. Yet, belonging at Lewis Island does not end when one crosses the bridge back to the mainland. Rather, belonging is maintained through discourse in person, by phone, by e-mail, and through letter writing and holiday cards. With or without direct reference to or presence on the island, people who "belong" at the fishery still belong *together* due to the shared values expressed in the Big Stories of tradition, the environment, and civility, while other commonalities arise through further contact, such as island maintenance during off seasons, attending particular community fundraisers and social events, appearing at the same viewings and funerals, and looking for one another in the Memorial Day, Halloween, and Winter Festival parades whether from the street or from the sidewalk.

In *The Sociology of Community Connections*, John G. Bruhn (2005) describes essentially the same phenomenon that Lattanzi Shutika sees, but favors a different key term, "social cohesion":

> Social cohesion is created; it is what happens when people share beliefs and values. Sharing a common purpose or destiny creates a bond between members of a community. Members invest in, and feel a sense of responsibility for, each other's welfare. An essential factor necessary to create social cohesion is stability. (Bruhn 2005, 206)

Bruhn's description fits crew membership particularly well, for even though many of the members of the crew see each other rarely outside of fishing season, crew members work together day in and day out within the season, and fishing happens within the context of more than a century of operation at that particular place. Moreover, when crew members do have a chance meeting at the grocery store, for example, the closeness of the crew connection is in operation, and fishing usually comes up in discussion. Crew members have an obvious "common purpose"—to catch fish as part of a team—and this expands with the larger purposes of preserving the tradition, protecting the natural environment, and building community. The responsibility toward each other is enacted during the fishing season in part because of the minimum number of crew members needed to do a haul. On a day-to-day basis, crew members are therefore aware of one another's work schedules, community activities, family obligations, and health. In a pinch, some members may forego or delay an outside obligation if the needs of other crew members have priority.

While Bruhn's description of social cohesion and Lattanzi Shutika's description of belonging clearly fit the Lewis Fishery crew, they also pertain to other people who belong at Lewis Island: the extended Lewis family, of course, but also the customers and visitors. Framed by a long history and shared concerns for the environment, the tradition, and a sense of civility, belonging at the Lewis Fishery operates in part because of the specific nature of the shared concerns: a concern for tradition obligates adherents to persistently participate over time, and a concern for civility encourages an atmosphere of respect and kindness that people will want to be a part of. While this dynamic seems to set up perfect conditions for belonging, belonging is neither simple nor uniform. There is significantly more inclusion at Lewis Island than people expect of private property, but there is not absolute inclusivity always. Whether and the degree to which one belongs stem from a combination of dedication, longevity, and physically being there on the one hand, and kindness, civility, and respect on the other. Practical circumstances may affect longevity, but commitment also comes from within, just as civil behavior stems from individual choices about how to act. The complex interaction of individual circumstances and choices across the settings of time and place determines belonging. An individual's placement on the spectrum of more or less belonging in relation to role (family, crew, customer, visitor) affects and is affected by narrative stewardship, as will be shown below in discussion of various roles and various populations at the Lewis Fishery.

As with any other family in the United States, membership in the extended Lewis family happens according to what anthropologists call consanguine (blood) and affinal (marriage) relationships, as well as legal and social relationships related to foster parenting. Being related makes it more likely that a member of the family will be involved in the fishery, but whether a family member is a crew member is a special consideration, following many of the same patterns whereby non-family members become crew members. Being a crew member differs from other nonfamilial roles at the Lewis Fishery (e.g., customer, visitor) for two main reasons: the considerable commitment required and the role of representing the fishery to others. Ted Kroemmelbein's description of the work commitment spans a variety of tasks:

> [There's] the amount of labor involved, collectively, to rebuild the bridge, tear the old one down, get in there on your days off, and your free time, and help out. There's a big commitment to fishing. It's not all fun and games hot dogs, and once a year at the end of the season, you know. [Diane Kroemmelbein: yeah, yeahhhh.] There's a lot of other stuff in there. (June 15, 2013)

Ted refers to some of the "perks" of crew membership, such as the end-of-season picnic and other social events, but makes it clear that the work is considerable and ongoing, and his wife agrees, having seen the effect of crew membership on both Ted's and her father's lives, as well as the lives of their family members.

In addition to the time commitment, Ted articulates the required commitment to civil, even polite, behavior that crew members might not uphold in other settings, even with one another.

> You know, one of the things about the island is you're in the public; you're with a group of people, so your behavior should reflect that, in that you don't, you know, run around swearing nonstop; you behave in a civilized manner, because you're a representative for everybody else there, so … and I think everybody pretty much adheres to that. No matter what we would say when we're with two of our buddies by ourselves, you don't say it when you're with Charlie and her kids [Charlie

Groth laughs. Diane Kroemmelbein: yeah] and anybody else who might be listening or whatever, you know [Diane Kroemmelbein: yeah], that there's different standards for where you're at, and then that needs to hold true there. Also, you're a representative for the island, so, when you're talking to people, you're polite, you answer the questions, no matter how tedious and boring [Charlie laughs] things may be or stupid, that is sometimes the occasion, but you're a representative, so . . . there's a responsibility that goes along with that. (June 15, 2013)

In this description, Ted expresses himself with humor (talking to me in a joking tone and referring to me in the third person; referring to the questions as "tedious," "boring," and "stupid" at times), but also speaks quite seriously about the responsibility of representing the fishery and living by its code of appropriate behavior. His direct mention of me does not indicate that as a researcher I'm missing the gritty behavior—I've witnessed the swearing plenty of times among the crew when children aren't present or in other places when we're together—but here I'm in the "mom" context in his description about civil speech standards in front of families. This concern for propriety never verges on prissiness, however, because the speakers are wearing stained and holey fishing clothes newly covered with fish scales, slime, and algae, often in the company of peeing dogs. Indeed, part of the crew members' commitment takes the form of being willing to be more wet, smelly, and dirty than is appropriate in other venues. In the same vein, Sarah Baker remarks that the success of her friend Matt Evans in becoming a full crew member when some other friends did not "stick" relates to him being more willing to "get his hands dirty" than her other friends (September 13, 2012). Despite physical filth, however, the "clean" tone Fred Lewis set of little-to-no drinking, swearing, or fighting has been maintained by the current generation of captain and crew, both for the sake of the women and children present (as Ted indicates by referring to my children and me, present and working) as well as for the sake of the "public."

Ted also introduces the topic of crew members representing the fishery through telling the story of the fishery. Crew member Dan Tuft links his experience of narrative stewardship to the annual Shad Fest:

That's usually when I feel like, oh, I know more about what happens than I think I do, because somebody will always be stopping you to ask something about the haul, or the river or the fish or that sort of thing. So there's a lot more curious people wanting to know about the whole fishing process. How long have you been doing it? And some people have a lot of questions. So, that's one of the few times I feel like I'm actually maybe sharing some of the story of the fishery, 'cause usually there isn't that kind of opportunity in a regular—not so much at least that I've noticed or that I've been part of, but at Shad Fest there is. (October 3, 2009)

When questioned further, Dan Tuft described "the story of the fishery" as "just the, what happens in a haul, or any of the history of the fishery that I can remember" (October 3, 2009). Pointing to the importance of narrative in Lewis Fishery culture, Dan identifies the haul itself as a "story" and links that to the fishery's history, which becomes the Big Story of tradition in our analysis here. Both Pam Baker (October 4, 2009) and her daughter Sarah Baker (September 13, 2012) refer to telling the fishery's story as a function of being a family member and a crew member—and as something of a rite of passage. Along the trajectory of ever greater belonging, Sarah fills the role of representative of the fishery as part of the crew and the family, sometimes with heightened responsibility, such as the time she spoke to the governor at the Shad Fest. Sarah enacts belonging: she, the story, the family, and the fishery all belong together.

The media and other visitors to the island typically ask which members of the crew are also extended Lewis family members, and some connection with the family has usually been the primary path to fishery crew membership. For Sarah, the practical connection is being *part* of the Lewis family, but there are other family-derived connections. Some people become connected with the fishery because their relatives are crew members—as was the case with Ted Kroemmelbein being introduced through his father-in-law, Johnny Isler, the Stout brothers on the crew of the 1930s, and some of the kids who help fish regularly. More typically, someone will have a preexisting connection with the Lewis family, such as the many coworkers at the Union Paper Mill who joined Bill Lewis Sr.'s crew and then later Fred Lewis's crew. Tim Genthner, a crew member from the late 1990s to 2013, married a woman from Fred and Nell Lewis's church who was a special friend of theirs despite being a few decades younger than the couple. Other crew members have come to it through friendship and community networks, such as the Marriott boys, who grew up a few doors down from the Meserve family, and Ken Bacorn, who started fishing when he and Steve attended high school together, then continued fishing when Steve left the state to attend college. Ken shows the pattern of multigenerational shad fishing families, too, for he is likely related to the Bachornes of early-twentieth-century crews, despite the different spelling of the last name. Moreover, Ken's son, Brandon Bacorn, was recruited during high school by *his* friend, Sarah Baker. In a friendship parallel to that of his father and Sarah's Uncle Steve a generation earlier, Brandon fishes with the crew even in years when his father does not and in 2017 began holding land, like his father before him. Occasionally, others who have a special interest in fishing will approach the fishery and volunteer, but often some other connection to the place and family is more significant, in part because duties at the fishery can compete with fishermen's time for other kinds of fishing.

It's important to note that it's not so much that you have to "know someone" in a nepotistic sense in order to be allowed to fish on the crew, but rather the considerable time commitment and messy, uncomfortable work makes the Lewis Fishery family reluctant to impose on just anybody, particularly now that they have no monetary incentive to offer. They also need to trust that the person they invite in is likely to "work out" in the sense that they will fit in with the crew as a group of civil and easygoing people, and in the sense that the fishery commitment could fit into their lives. This same situation applies to friends of current crew members as well. When the fishery family and other crew members brainstorm possible new recruits, they consider people who live and work close enough to make participation on a weeknight possible, then consider other factors that might make someone a good candidate. Since the mid-1990s, I have seen a couple dozen people come fishing for part of or more than one season. Responding to the example of a man who constantly joked about payment and did not last more than one season, Muriel Meserve says, "No, it weeds out itself very quickly on that, because it's an awful lot of hard work and no remuneration for it, so, you have to really [laughs] believe in what you're doing ... and really like it. Otherwise, the rest of the people tend to drop by the wayside then real fast" (October 5, 2009). No matter whether they come to the fishery through person-to-person connections or personal interest, belonging and longevity grow through a combination of practical considerations, commitment to at least one of the Big Stories ("You have to really believe in what you're doing"), and a certain degree of amiability—the same basic elements the Lewis family considers when recruiting.

The issue of civility has been addressed earlier, but the quality of being "a good sport" bears more discussion here. Being willing to be wet and filthy is part of it, but during the past half century, being willing to go through the misery for no return for long stretches of time has become a necessary part of a crew member's mettle. When Pam Baker calls the crew the "village idiots" for dedicating so much time to unsuccessful fishing in bad weather, she adds, "But you have to keep *going* or else *nobody's* going to know about it" (October 4, 2009). Here, Pam connects the misery of the fishing to an obligation to bear witness. The Lewis Fishery needs to keep fishing to show that the fish aren't there, then figure out why they are not catching.

However, it's not enough to fish and see; one must also tell the story. The commitment required of crew members entails both work and narrative stewardship. Pam Baker recalls a short-term crew member, Greg Dunaj, who "was very into the history of it, and the community of it" and wrote beautiful blog postings that season (October 4, 2009), but eventually had to drop off when a job took him elsewhere. In his blog about the 2009 season, Greg says of his first night rowing (a big step for a new crew member), "Truthfully it is an honor to be asked to row; I had felt like a stranger in the town I had lived in for the past 20 years. To be accepted like this for such an integral part of the crew is a great thing and I am pleased" (Dunaj 2009). In this and other posts, Dunaj echoes the appreciative attitudes expressed by other crew members. Dan Tuft refers to being part of the Lewis Fishery contingent at the Smithsonian Folklife Festival as an "honor" (October 3, 2009), while Diane Kroemmelbein remembers that, being "little kids growing up on the island," "you felt like you were part of something big, like big, you know how when you're a little kid you use the word 'big,'" in part because the Philadelphia television networks would cover the event (June 15, 2013). After commenting on how important his relationships are with the crew members of all ages, Diane's husband, Ted, ironically draws together the misery of fishing and his appreciation:

> I really feel privileged to be able to do this, and I thoroughly enjoy it. I enjoy the being cold. I enjoy the being hot. I enjoy the being wet. I enjoy being stinky, full of mud, away from my family [Diane Kroemmelbein laughs], you know, all the stuff that are the downsides, but in the end, it's all a plus. It's the opportunity to do this. I don't know if I was born for it, but I was born ready for it. (June 15, 2013)

Ted underscores an important point here: the fishery crew members are not necessarily drawn to the fishing and the fish so much as they are drawn to being outdoors with other people doing something that is increasingly rare in today's world. For Ted, the unique combination of practical situation, values, and temperament has fallen into place.

While the crew members and family are the people who most "belong" at the fishery, at the other end of the continuum are the media and those who visit *only* during the Shad Fest. In general, these groups do not put in the time at the island that others do. Moreover, their motivation for being at the island is quite different: they come primarily for their own entertainment or to do a task related to a profession that's generally *un*related to the island and the fishery. In the case of the media, the thin connection is in some ways intentional. Newspaper photographer Mary Iuvone, for example, reveals that "when we were out on the boat . . . I just wanted to be, you know, just a fly on the wall, just sort of, you know, I didn't want [Steve] to even know that I was there, you know, just let him do his thing, and I would just photograph it, him doing, and everyone else just doing what they're doing" (May 20, 2011). In her desire to get the best picture, she pursued disconnection in a context in which connection is a central value, at that moment showing belonging instead to her profession governed by journalistic objectivity. Despite the fact that the media operates in a liminal space between connection and disconnection, these professionals play an essential role in the system of narrative stewardship, focusing on the Big Stories (Groth 2011). Keziah Groth-Tuft describes a heightened energy when reporters are present. She says, "There's usually more stories told then, 'cause like Steve's telling them stories so then you get to hear more stories too, and you can ask later. You got to be careful what you say . . . you just gotta make sure like what you say can't be taken the wrong way" (October 3, 2009). As a middle schooler, she already understood the crew's important responsibility as fishery representatives, and she realizes the media's role in narrative stewardship.

Keziah, Ted, Dan, and other crew members also comment on the need to tell stories directly to the public when Shad Fest visitors come to the island during the haul. The event generates a certain amount of tension. On the positive side, the family and crew appreciate the opportunity to share the island and the Big Stories in a process that might draw people back to the island when it is quieter, to create more meaningful connections and understandings. On the other side, being on their best behavior and on stage can be "work" for the family and crew, especially when some of the festivalgoers are not on their best behavior. Some festival visitors have arrived on the island loud and tipsy, booed the crew if no fish are caught, shown disrespect for property (e.g., climbing on other people's docks attached to the island, causing unnecessary stress on an ailing bridge by jumping up and down to feel it bounce, trampling plants, throwing much-needed stones into the river), and even ridiculed the tradition. The policing role alone can be wearing on a group that dislikes conflict. The following exchange between Ted and Diane Kroemmelbein and Kelly McMichael shows a certain note of "it is what it is" acceptance:

> TED: You know, for the most part they're just tourists. They want to see it. It's like, I don't know, the same people that go see the fireworks. OK, it's available, so let's go down and see it, you know. . . . They're going to haul shad; let's go watch them do that, you know, one time. I don't know, I don't think it's any real interest, it's just people [who] are interested in seeing what we do—
>
> DIANE: People bring their kids to show them a little bit of what's happening outside their own world.

KELLY: But they don't even know what they're looking at.
DIANE: Yeah.
TED: But that's okay.
DIANE: Yeah, it is okay.

Shad Fest interactions with the general public go into a different category from most other interactions with visitors on the island. Given the nature of the event and the size of the crowd, these interactions are more like a performance than a neighborly conversation. Over the years, several fishery members have drawn on their collective experience in theater and teaching to employ more crowd-management techniques. Caution tape and sawhorses communicate appropriate boundaries while bagging up, and a rope across the stairs prevents festivalgoers from walking in on the crew changing clothes in the crew's cabin. The demonstration hauls themselves operate more like performing arts productions, with the captain giving a little speech and answering questions while the crew loads the net onto the boat. Here, my role has grown to take advantage of my research and teaching skills. I have not done much of my usual crew work during the Shad Fest since the mid-2000s, but rather stayed down with the crowd to explain the process and answer questions before the visible part of the haul happens. While the Shad Fest disrupts the usual workflow and represents another commitment for the crew, it also allows the fishery to tell those who do care "what they're looking at" and inform interested people that they can come back to the island to watch a haul any night during the season and can enjoy the island any time the gate is open.

The more managed demonstration has made the narrative processes during the Shad Fest more complex, with stories told to the crowd at large, stories told to especially interested day visitors, and stories told "backstage" within the crew and between the crew and community members. Keziah says,

> You tell a lot of stories about other Shad Fests. And you tell stories about like dumbos in the crowd [Charlie Groth laughs], like this person asked this and this person was a *jerk*. Stories come out once you get up to the top, and there's no more people there, because like, you tell a lot of stories like *to* the crowd, when you're down there? When you get back up, it's like you're alone again, which is nice, I like that part. (October 3, 2009)

In the first part of this statement, we see the teen's enjoyment of the ironic difference between front stage and back stage, and the enjoyment of recalling other Shad Fests. Yet, we also see at work narrative stewardship regarding behavior: emerging microlegends about "jerks" in the service of the Big Story about civility. More than taking part in entertaining or in cruel gossip, the kids and adults reinforce the expected behavior on the island with the negative examples of anonymous "dumbos" and "jerks." Badly behaving festivalgoers are also compared with other tourists who come to the Lambertville–New Hope area from New York City throughout the year, some of whom are known for their demanding and disrespectful behavior in the classic tourist-townie dynamic. The "jerks at Shad Fest" stories form part of the surveillance subgenre of microlegends. While the crew might talk about them again during the week after the Shad Fest, it's important to note that bad behavior stories will be interspersed and contrasted with stories about the really nice, interesting, and interested people at the Shad Fest, some of whom visit again in the two weeks following the festival and start to move further along the belonging trajectory, making the transition from Shad Fest visitors to regular visitors to the island.

Between the Shad Fest visitors on the one hand and the family and crew on other, in the shifting middle ground of belongingness are the customers and people who visit the island at times other than the festival. On any given night, a combination of visitors appears, usually including friends of the family and crew, people from the area, dog owners, customers, and people who have heard about the fishery through word of mouth, the media, or a Shad Fest visit, and are coming to see the fishing, see the crew and family, or both. In a brief description of the mix of people and relationships during a night of fishing, Keziah says,

> Well, I guess there's . . . Mr. Van Hise comes down, and Robert used to come down, people who were on the fishery [crew] before. . . . They come down and we'll make sure . . . they're doing okay. You let them come up to the cabin if it's cold, depending on how good you know them and stuff. And when we sell the fish and we make sure everyone gets it, and our better customers and stuff that are like really important. (October 3, 2009)

Morgan Van Hise is a customer from a nearby town who came to the fishery for decades until the last year or two before he died, in 2017. Robert Johnston was a crew member for decades; he retired in 1995 but would periodically visit the fishery until his death in 2009. In cases where long-term visitors visit occasionally, everyday storying will quickly commence to "make sure they're doing okay." Keziah mentions here two other important milestones that show belonging: being invited into the crew's cabin, a private area, and getting special attention as a "better customer." While all the customers are important, and the fishwives try to spread the fish around fairly, when the distribution must be unequal because of a discrepancy between available fish and customer needs, "better customers" will be favored. As discussed elsewhere, the amount of money spent might play some role in defining a "good" customer, but this is less important than longevity, willingness to come to the island, and, most of all, friendliness and understanding the difficulty of distributing a small catch.

Interaction with the island's "space" is another way people negotiate belonging on the island, as Keziah notes when she refers to some people being invited into the crew's cabin. This distinction points out that while everyone has the potential to belong on the island, not everyone will belong everywhere on the island: the house, its fenced yard, and up in the crew's cabin remain private. Usually the only people who are invited into the cabin besides the crew and family are people who are volunteering on the crew on a particular night and people who especially need shelter from the weather. *Under* the cabin is not private space, but not everyone chooses to belong in the fishery "market." Customers come under the cabin, of course, but so do people particularly interested in the *whole* process, such as parents with children. These folks not only become associated with that additional space at the fishery but also interact more with the fishery women, taking part in more everyday storying and storying about fishery traditions related to procuring and cooking shad. Similarly, those who walk dogs or take meditative walks on the island are more likely to walk past the private house and yard, then go up the more intimate—but nevertheless *public*—center path to the garden. These visitors sometimes notice and report problems on the island, becoming part of the surveillance team and the related narrative activity. Thus, the extent of physical presence and storytelling together create one's "belonging" at Lewis Island, and for visitors and customers especially, this relationship varies greatly depending on the individual and over time.

Perhaps most instructive is to consider who definitely *doesn't* belong on Lewis Island, and the simple answer is, people who do not behave well. One can be aloof and not really "belong" on the island, but a tenuous connection does not preclude coming to belong over time. In extremely rare cases, however, people will be told they cannot return, and it is always because they have repeatedly done things that violate the island or its code of civility, evidenced by the surveillance microlegends discussed in chapter 4. Even then, however, the Lewis family is known for demonstrating a great deal of patience and restraint in sharing the island.

While physical presence is a big part of belonging, it is not essential at all times, because people belong again as soon as they cross the bridge after a long absence; even in death, people still belong to the island. Much like the conversations Mary Hufford (1992) describes in which Jersey foxhunters tell stories to bring now-deceased generations of friends back into the social scene, Big Stories and microlegends bring family, crew members, customers, and visitors back to Lewis Island in conversations that enact their belonging, even after death. They then become part of the "narrative texture" (Johnstone 1990), "thickening" (Casey 1993), or "invisible landscape" (Ryden 1993) that is essential to Lewis Island, belonging more than ever to the place, a place to which others they have never met will also come to belong.

Having established that belonging at the fishery stems mostly from particular roles, circumstances, and choices that govern presence, behavior, and storytelling, I turn now to belonging in the sense of social and cultural diversity. Again, the picture is not simple. Like any small town, Lambertville can seem a bit closed to newcomers at first. One neighbor born and raised elsewhere shared that she was told she would be a "drifter" until she had lived in town for twenty-five years, at which time she would become a "newcomer" (Janine MacGregor, personal communication, September 6, 2015). On the other hand, there seems to be greater acceptance of people who would be excluded elsewhere.

For example, Pam Baker tells the story of a woman who bore two children out of wedlock in the 1930s and was kicked out of college a few months before graduation when she became pregnant with her eldest. However, she got a job at a local business, which in turn created a job for one of her children who had cognitive disabilities and whom her extended family and the town looked after long after her death. The woman and the children's father remained unmarried and lived separately, although the father clearly loved the children, and the community supported the mother in various ways as she raised her kids. Pam recalls that although she knew all the players, it wasn't until she was in high school that she learned about the unorthodox arrangement, because no one had made a scandal of it and Pam was raised to simply accept people and their situation: "I think about that now—even *now* that's a little quirky. I mean, b-but [the] *thirties?!* Right, right, so I mean this town must have been fairly ... forgiving" (October 4, 2009).

Lambertville is also more accepting than many other places of other types of family diversity. While in many ways the traditionally all-American town where kids grow up, marry each other, and raise families in conventional ways, Lambertville has long accepted mixed-race couples and same-sex couples—in fact, Lambertville's mayor was the first mayor in New Jersey to perform a civil union ceremony, and later a same-sex marriage ceremony. Moreover, there is marked acceptance of people with a wide range of abilities.

When I first came to live in Lambertville, having grown up in a nearby town, I was struck by the number of people who spent much of their days walking around town or sitting on park benches or stoops in nice weather, many of them apparently with cognitive, physical, or mental health difficulties. Many of these folks are "looked after" by various people in town, officially or unofficially, and the town at large accepts their presence. Loitering charges are not threatened, and people with a wide range of abilities are included in the community life of the downtown, the churches, and the library.

Even in a recently gentrified town, the Lewis family's connections with the mills (associated with the older members of town), the performing arts (associated with many newcomers), and education (essential to everyone) break down some of the barriers. When Ted Kroemmelbein describes the fishery community, he describes a community linked to place through tradition, but where the people are not homogeneous in terms of age or socioeconomic group:

> The diversity is what, what I like. You know, there's professional people; there's working people; there's old people. You know, now we've got a couple old guys in the last couple years now; what're they like seventy-five, eighty? Couple of these guys come in and bring boots? Just to pull the net, you know? To help out, which I think is so neat. Like this year also we had a really good crop of young guys. And I think we need that more than anything else, 'cause we're starting to get a little long in the tooth here. (June 15, 2013)

Ted starts out by describing occupational/class diversity, which also relates to diversity of education and bridges differences between the old, mill-centered Lambertville and the new, gentrified Lambertville. He moves on to describe an age spread of men wider than what one usually finds in other workplaces (see below in this chapter for a discussion of the increased number of children on the crew).

Although Ted does not mention it, I would add that the crew members also differ in politics and religion. Political party membership is rarely if ever discussed, but the group has included Democrats, Republicans, and independents, as might be expected in a business rooted in two blue (in the sense of mostly Democratic) towns within red (mostly Republican) counties. Liberal and libertarian ideas coexist in an atmosphere that values integrity and kindness. Political issues and current events are discussed with skepticism for political competition and spin, and at the same time an ideal of faithful and responsible citizenship and governance as service. Voting in local elections is encouraged, such as the school board and fire chief elections, which might occur during fishing season. As in many families with mixed political alignments, crew members generally avoid discussing partisan politics but might take up the topic in smaller groups with similar views, after the haul or at a different location.

Similarly, religious and spiritual identities vary and, in some cases, affect crew members' motivations in volunteering. Largely reflecting the white European American mainstream in this region, the crew and their extended families include mainly Catholic and Protestant Christians but also Jews, Unitarian Universalists, and religiously unaffiliated people with various beliefs. A number of the fishery community are humanists, agnostics, and atheists who may or may not belong to churches. Many crew members are church leaders, and church-related obligations may affect the crew's makeup on any given night, as may the Easter and Passover events that fall during the season. Moreover, many crew families are heavily involved in Girl Scouting, a movement and organization that is at base spiritual and values based, as represented by its Promise and Law, but that does not define specific spiritual or religious beliefs members must adhere to, even allowing members to substitute other wording for the word "God" in the Girl Scout Promise, as traditionally written (Girl Scouts of the United States of America 2018, 24). Thus, religious diversity and moral character, concerns shared by the Girl Scouts and the Lewis Fishery, reinforce one another. At the fishery, as with political affiliations, religious and spiritual differences are largely downplayed while commonalities (the work of religious communities; the values of kindness and integrity) are shared openly. In short, religious and political allegiances are varied, and at the fishery they take a back seat to belonging and the Big Story of civility.

The largest change in *crew* diversity is that women and children became essential members of the fishery crew over the course of the late twentieth century, even though adult males still hold prominence in the traditional activity.[1] When William Lewis Sr. founded the Island (now Lewis) Fishery, he owned it and led the crew as well as selling the fish on the bank. With the next generation in the first half of the twentieth century, women began to run the sales, and today males—even the captain on a short-staffed night—are usually only "helpers" under the cabin. Females becoming full members of the fishing crew, however, has taken much longer, influenced by both individual biographies and wider societal and natural changes.

As a teenager in the 1970s, Sue Lewis Garczynski was the first female allowed to row with the men during the haul, a privilege hard won through arguing with her father, Fred Lewis. Both Sue and Muriel had the example of their mother, Nell Lewis, breaking gender barriers as the only girl competing (much to *her* mother's chagrin) in a 1920s marble contest. However, because

Muriel was seventeen years older than her sister, only Sue had the advantage of some societal changes that promote gender equity, such as federal Title IX legislation. It took over another decade after Sue Lewis rowed with the crew for Sue Meserve (Steve Meserve's wife) to become the first female to earn a full share as a crew member, not only reflecting the considerable work she put in but also due in large part to Steve Meserve's advocacy with his grandfather. Years later, I became the second female crew member to earn a full share, and the first woman to hold land. Gender barriers have been broken on Lewis Island as in wider US society, and on some nights there are more females than males on the crew (humorously referred to as a "chick haul"). Yet, as in wider society, which has not yet achieved full gender equity, there is still a gender hierarchy with males almost always first (and second, and third) in the lineup for rower, landsman, and—if they want it—pulling first leads. This practice is in part a matter of physical strength (on average, the male adults are stronger than the female adults and children), but there is also a social hierarchy, as evidenced by the fact that arguably the best lead pullers have been two adolescent girls and one adolescent boy. Ultimately, today the fishery operation would be hard pressed to operate solely with male youths and adults as it did when Bill Lewis Sr. was captain, because of natural and economic reasons.

Children have been present at the fishery since Fred Lewis was a small boy, as his widower father brought him down to the fishery and allowed him to perform kids jobs like placing the buckets and working the cork line. Now, however, when most mothers must work outside the home—as well as at the fishery by choice—it is much more commonplace for children too young to stay home alone to work at the fishery alongside their mothers, fathers, and grandfathers. At the same time, the closing of area mills and the drop in fishing profits (with the drop in shad populations) have led to a smaller pool of available adult males from which to recruit crew members. Ironically, although the increasing gender equity trajectory has allowed women to join the crew, gender inequity is part of keeping them there: now that crew membership is a fully volunteer endeavor, the work more resembles the unpaid labor women have contributed for generations elsewhere in the communities. Nevertheless, the work of women and children has become essential for the operation of the fishery.

Despite the obviously diversifying age and gender makeup of the crew, outsiders are reluctant to recognize women's increasing role at the fishery. Instead, they promote a view in which only males belong. Renée Kiriluk-Hill notes that her newspaper looks for "the traditional picture of the crew" (June 3, 2011), which is the crew all standing together looking at the camera. In a picture taken in 1937 for the *Newark Sunday Call*, all the crew members are men. In 1970s crew pictures belonging to the family, Sue Lewis is the only female pictured, but in 2004's Shad Fest snapshot, six females out of thirteen crew members are pictured. Many visitors have trouble accepting the crew's gender diversity. Even when female crew members stand clad in hip boots, visitors ask the women, "Are they [the men] going out fishing?" when it's obvious they should be asking, "Are *you* going out fishing?" Muriel Meserve notes that "even if they see you [the women] pulling the nets in and everything, they still will ask the guys about the fish" (October 5, 2009), and concludes with a laugh, "That has taken longer than anything to change" (October 5, 2009). The media is often part of the sexism: one time when Sue and I were in position to perform the essential function of catching the boat when it landed, a photographer went so far as to ask us to move so that he could get a better picture of the men in the boat, rather than include us in the frame as part of the operation.

In this complex and mottled assemblage of gender equity and inequity on the Lewis Fishery crew, Steve Meserve remains an advocate for fostering women's "belonging." For example, after the Saturday haul of the 2014 Shad Fest, a photographer from a local paper asked for a picture "of all the guys" lined up around the fish the crew just caught. As I began to step out of the picture and the men began to line up for the photo, Steve and Pam called the women and girls by name, asking us to come forward and get into the picture, contrary to the photographer's request. Later, Sue Meserve and Pam Baker both recalled the event with an air of being offended by the photographer, but being grateful to Steve for his sensitivity and willingness to counter the media's male bias. Just a couple of weeks later, Steve made the unusual decision to allow me to row in place of a young man on the crew who rowed in regular rotation with the other men. When the young man expressed surprise, Steve told him that I had rowed before when the young man was not there and noted that with a high river, it was a situation I could more easily handle, which was true (I'm *not* as strong as this young man, and my back is lousy), and so I would have a turn. In the same way that Steve carefully shows interest in all the male crew members to make sure they continue to enjoy the volunteer work, he made choices to show that I, too, belonged.

The most important effect of this age and gender diversification is that the fishery continues past its 125th season and into the family's fifth generation, a process that parallels the ethnic diversification of the customer base. Categorical sameness does not make or break belonging on Lewis Island, to be sure, and although the crew is diverse in the sense of gender, age, occupation, education, politics, and religion, it is rather homogeneous in terms of race and ethnicity. As with the increase in female crew members, individual biographies and wider social trends bring people other than white European Americans to Lewis Island, and the fishery community supports inclusion as part of the Big Story of civility and kindness.

Other than the ethnic mainstream, Jews were perhaps the first ethnic group whose group presence was felt on Lewis Island. The Lambertville–New Hope area did not have its own synagogue until Kehilat HaNahar (Little Shul by the River), a liberal Reconstructionist shul in keeping with the community's liberal nature, was founded in 1994 (Kehilat HaNahar 2015), and a Jewish *community* coalesced. Before that, Lambertville had a small Jewish population of individual families, including that of Joseph Finkle, a Russian immigrant who founded Finkles Hardware (Castagna 1999–2002), a regionally known Lambertville institution for generations (in fact, the fishery's account at Finkles is still in Fred Lewis's name a decade after his death). Frances Finkle, Joseph's daughter-in-law and a founding member of Kehilat HaNahar, moved to Lambertville from Trenton in 1961, when she married into the Finkle family. She recalls that there was no sense of a Jewish community in Lambertville then (personal communication, July 30, 2015). The Finkle family went elsewhere to observe the high holidays, and Frances Finkle had a part in the decision to open Finkles Hardware on the second day of Rosh Hashanah, contrary to common Jewish religious practice at the time, to meet the needs of their gentile customers.

The Lewis Fishery's business practice mirrors Finkles' acceptance of difference. Despite the small Jewish population *in* Lambertville, in the early twentieth century the Lewis Fishery supplied fish to Jewish communities in Trenton and other places farther away. Fred Lewis had a Jewish clientele large enough that he successfully lobbied the state to legally change the fishing season so that he could provide live fish that could be ritually prepared for the Jewish winter holidays, according to Jewish law.

The other significant turn in the ethnic history of Lewis Island came in the late twentieth century, when many Asian immigrants of various ethnicities settled in central New Jersey, learned about the Lewis Fishery, and integrated into the customer base. Pam Baker shows appreciation for the diversity of the clientele and their love for the particular kinds of fish the fishery provides:

> These people who come, like, the different *Indian*, the Indian population that's been coming for years and years and years, *decades*, because they *love* it. And the *African American* population, who likes, you know, the catfish, and it's a great thing. And we are a very diverse ... *community*, and people don't *see* that, so it's so nice. *We see it* and I don't think that a lot of people appreciate that there *is* a

diverse community in this area and that we *are* [diverse], and it's just so cool. (October 4, 2009)

Pam's positive reaction is echoed by other members of the family and crew, supporting the idea that welcoming cultural and racial diversity is part of the Big Story of civility. In fact, the women on the crew in hip boots and tank tops standing next to female customers in saris or niqabs make a stunning contrast and point to the specialness of this place within the wider area. Other differences are more subtle.

When initially drafting this chapter, I was discussing with Keziah my work about differences between crew members, customers, and other visitors, and she added quickly, "Yeah, the crew doesn't eat the fish" (October 3, 2014). Her keen insight was startling, although I must make the distinction that the key factor isn't that the crew *doesn't* eat the fish, but that the customers *do*. The crew are not drawn by eating the fish or even fishing per se, but mostly by fishing in this way in this community. None of the crew members eat a steady diet of shad, and most have just two or fewer shad meals in a year. The African American population is drawn by a combination of the fresh fish and generations of connection to the Delaware River. In contrast, Asian customers, including people from India, Pakistan, Bangladesh, and China, are drawn *only* by the fish supply. Further, cross-cultural interactions as well as foodways differ. Narrative activity between the white European American fishery people and the African Americans emphasizes "getting-to-know-you" on the individual level, including touchstone narratives, because for generations these groups have shared a first language and a regional culture. Everyday storying between the fishery people and Asian customers, on the other hand, sometimes faces a language barrier and includes more cross-cultural exchange, in addition to making connections as individuals.

Like Fred Lewis's changes to accommodate the needs of the Jewish population, the process of getting to know the different Asian ethnic groups has impacted the way the fishery operates. Over and above knowing that Chinese customers are more likely to want shad heads (even without the bodies) and suckers, the women selling fish have had to consider the very nature of the selling process to create a common ground that follows the fishery family's values, yet answers the needs of Asian customers. The Lewis Fishery remains adamant about having as wide and equitable a distribution as possible, giving small advantages to those at the top of the list and those who are nice to others, and charging a flat fee, although they occasionally offer a discount in unusual circumstances.

Over time, the sales method has gotten more complex and diverse as more ethnic groups have come to belong to the fishery customer base. In particular, the fishery's attempt to distribute shad as equitably as possible with flat fees has, at times, clashed with some Asian traditions of negotiating or "haggling over" the sizes and prices of fish, with each person expected to pursue the best deal. The sellers have come to understand the cultural difference and not take offence. Instead, they work to communicate how sales work at the fishery. In situations when two customers from haggling traditions are competing with each other, Sue Meserve might divide a heap of fish between the two at the same time, picking up similar-sized fish in pairs and sorting them into two buckets to make it obvious that distribution will be as equal as possible. Usually, after a few visits, new customers from these cultural traditions become familiar with the flat-fee and shared-distribution customs of the fishery. Yet, on one occasion, I witnessed a customer getting ejected from under the cabin for repeatedly insisting that he get all the fish he wanted—the preferred number, sex, and sizes—intending to leave other customers with nothing, no roe, or only the small fish.

In response to cooperative customers, on the other hand, for many years the Lewis Fishery tried relaxing the strict rule that one had to be at the haul to get shad. At the time, two Indian women faithfully arrived early, stayed late for a second haul, and allowed other customers further down the list to get something from the haul. Conversations revealed that the women were buying large numbers of fish, not just for themselves but for other members of their community to support traditional foodways. As trust grew between the two groups through storytelling and repeated smooth sales, an understanding developed that these customers would come to the island to be sure to get fish from the haul, but if they chanced it, they could stay home and wait for a cell phone call after the other customers were taken care of. If there were enough fish to make it worth it, the customers would either drive down to Lambertville, or they would negotiate a rendezvous later closer to the Meserves' and their own homes. After several years, however, arguments arose when, with an influx of more new customers and even more widespread cell phone use, several customers came to believe that if they phoned ahead, they were first on the list, even if others phoned ahead, too. In 2015, the requirement of being on site when the fish landed was resumed to limit conflict. In 2017, the sellers experimented with distributing two fish to each customer first, then going through the list to negotiate the remainder, again in an effort to distribute fish equitably among those who committed to being there during the haul.

The fishery family and crew have made efforts in addition to adjusting selling methods to build bridges across cultural difference, reflecting the Big

Story of civility and the actions of the individuals involved in other places. In the years since the attacks on New York's World Trade Center on September 11, 2001, and the increase in anti-Muslim sentiment in the United States, acceptance of Muslim customers from India, Pakistan, and Bangladesh on Lewis Island has gained heightened importance. Arabic is often heard on Lewis Island, because as the religious language of Islam, it is sometimes the common language of Muslim customers from different ethnic cultures, allowing customers to converse with each other more fluently than in English. In 2014, Keziah began slipping in conversational social phrases in Arabic, a language she'd been studying in college. Startled at first, these customers expressed great joy and surprise; so surprised was one man that he grabbed his head with both hands and stuttered, "But, but, how do you know this?!" For several years, the fishery women and a regular customer whose abaya and niqab covered all but her eyes would shyly exchange hellos and smiles, apparent in the Muslim woman's eyes. One night in 2017, while mending net, I noticed this woman cross the bridge from the island, then cross back over. A little later, I mentioned to her husband that I was sorry I did not get to say hello to her, and he told me that she had come over to say hello, but could not come down to where I stood mending net because I was in such close proximity to the men on the crew. Still later in the haul when the men were elsewhere, she and her husband returned, her husband introduced us, and she held out her hand. Over the next couple of nights, I took care to wait until the men were at a distance and then introduced her to the other women in the family and crew.

The growing connections across cultural difference became even more important in the 2017 season, as the anti-immigrant tenor of Donald Trump's successful presidential campaign the previous autumn resulted almost immediately in an increase in hate speech and vandalism in the Delaware Valley. During the Shad Fest, an especially large and diverse group of new customers competed for the catch under the cabin. Emotions were high, since it was the only haul we could sell during the festival, and, as a midafternoon haul, it had a smaller yield than an evening catch. One white woman pushed for attention before the rest, and when Sue Meserve told her that she had to be there when her name was called and that she would get her turn like everyone else, she walked out. On her way off the island, she complained to other crew members that the immigrants were taking all the fish, matching the exclusive line of thinking her husband had been pursuing with unsympathetic listeners outside the cabin. Pam Baker politely responded that she was sorry, but the Lewis Fishery needed to accommodate everyone, and if the fishery could not satisfy her, she was welcome to buy fish elsewhere. Under the cabin later, when the woman's turn came up, Sue called her name several times to sell her her share of the catch, but by that time she had left the island in anger. Later, when the crew shared their different experiences of the incident, it was unanimously agreed that it was better to lose a new customer who was intolerant of others than cater to bigotry and treat loyal customers disrespectfully.

Those present clearly found the incident disturbing, as it was a rupture in civil discourse and inclusion; one is reminded that social ideals are inevitably realized imperfectly at best. British sociologist Brian Alleyne recommends seeing community "as a network of agents with ever-changing projects rather than a tapestry of people with shared roots" (2002, 622). In this chapter, I have attempted to break down the now conventional and at times romantic concept of a tapestry of diversity into individuals and groups with a variety of "ever-changing projects," and biographical and cultural trajectories. One might be an immigrant looking for a food source to help continue the cultural narrative embedded in a traditional foodway, a visitor returning to the area using a touchstone as a tool of impression management (Goffman 1959) to reclaim local or regional identity, a member of the fishery family furthering the Big Stories of tradition and environment, a Girl Scout leader introducing girls to traditionally masculine activities, or a newcomer to town finding a place in the community.

The cumulative process of belonging—individually or as a group—is neither neatly patterned nor complete at any one time. Yet, narrative practice consistently plays instrumental and expressive roles in belonging on Lewis Island. Debra Lattanzi Shutika (2011) well describes how individual, everyday, social narrative behaviors build lasting meanings:

> The daily experiences that constitute a life are fleeting, yet humans are compelled to relive and recall the commonplace events through narrative exchanges with others. The stories one shares regarding personal events reconstitute individual experiences for the teller as well as for those who listen. In the process of creating a story of daily life, what was once fleeting gains a measure of durability. (2011, 80)

What she describes reminds one of everyday storying on Lewis Island, which strengthens interest and connection for those who choose to initiate or participate in the process of narrative stewardship as tellers, listeners, or both. Lattanzi Shutika's words also apply to other narrative types. Touchstones provide an opening for new acquaintances, then microlegends and character anecdotes pique interest, and Big Stories foreground values and meanings that resonate more widely. The patterns of the haul and processional storytelling function as instruments of belonging, for they invite all present to co-create stories, to commune through narrative.

EIGHT

"A Whole 'Nother Place": Narrative Stewardship and Sense of Place

It's a beautiful spring day in 2005, belying the danger we find on Lewis Island as we meet up with other members of the fishery family and crew. We have come to clean up after the third-largest Delaware River flood in recorded history. For years we have been teaching our children about the dangers of the river, but these lessons had more to do with life vests and the water's edge. Today, we see mud and debris everywhere left by the receding waters, but the greatest danger is microscopic. We give the kids a quick biology lesson about bacteria, wearing gloves, and disinfecting before eating—and don't touch your face absent-mindedly, or someone will need to come disinfect you. For years we have walked over the footbridge to this "whole 'nother place." This beloved place is special in its peace, kindness, and beauty. Vivien Cosner, a high school classmate of Captain Steve and a New Hope resident, calls it a "haven" (personal communication, December 14, 2017), and other island visitors would agree. Today, however, it's decidedly different as we row over Island Creek to the island, the bridge having washed out as expected, but with damage on the island much greater than usual. Our home place shifts, and so do our identities with it.

My work detail is in the house on the island, and I am disoriented as I walk to it, imagining the water that placed debris in the branches more than ten feet over my head. Today I'm working in the bathroom and adjacent pitch-black walk-through closet on the main floor, a story above ground level where the water crested at three inches above the windowsill, and now three inches of mud sits on the floors of kitchen, bedroom, everywhere. We need to shovel it up before the mud hardens. My sense of place turns topsy-turvy as I stand in the familiar living room and imagine the river, a natural force, pushing through the domestic place, planting its signature stain, engulfing most of the furniture but leaving unscathed the framed pictures hanging on the wall. Fumbling through the dim bathroom—we won't have electricity on the island again for the next two fishing seasons—I find on the vanity a small pile of keys, coins, a business card or two, a note with someone's name and phone number, a safety pin, a pocket knife. I recognize this miniature place with a rush of poignancy; my father always deposited a similar pile of miscellany on his bureau top each night, just like my husband does. Only days ago, all the men on the fishing crew collected similar small piles of coins from their pockets to assemble a prize for the first cleanup crew person to slip and fall in the stagnant muddy water; my daughter wins it and celebrates her luck. Here by the bathroom sink awash with river smell, this seemingly random collection is the contents of Fred Lewis's pockets the last night he spent in this house, before the flood, before he and his wife moved from the island to spend their last winter together at their daughter's house where they could see the island across the creek, before he died the previous spring. This small, countertop place recalls Fred's life on the island at an intersection between his intimate life and his position as a public figure, so recently memorialized across the region. I take a deep breath and recommit myself to the cleaning project, to the process of restoring the island to safety. Although it will never be the same again, the island will be returned to its usual peaceful nature, to be shared again with the community.

Years later when we remember the spate of big floods at the turn of the century, this will be one of the stories I tell. This time of dumpsters and work parties throughout town will change a couple generations' sense of place, even their identities. In periods of responding to floods, flooding becomes a larger part of Lewis Island's and Lambertville's sense of place. When Hurricane Katrina happens, Lambertvillians will understand better than most why the folks from New Orleans do not want to leave their homes; we are connected to them by our own flood stories, tactile memory, and sense of place.

Scholars and the general public have long considered sense of place, seeking to define what it is, how it develops, and where it is on the trajectory of evolution—assuming there is such a trajectory. After an extensive literature review, sociologists Mike Savage, Gaynor Bagnall, and Brian Longhurst succinctly conclude, "Over the past two centuries sociologists have frequently pronounced the end of local identities and yet attachment to place remains remarkably obdurate" (2005, 1). Sociologists perhaps began the lament for loss of sense of place earliest, because of their concern for social changes accompanying urban and industrial development, including the movement of people from rural areas and the inherently place-based agricultural occupations into increasingly urban, human-devised spaces. Anthropologists and folklorists, on the other hand, came to this discussion later, because of the former's focus on cultural groups outside the reach of industrialization and the latter's focus on preservation of older lifeways despite the creep of industrialization.

By the end of the twentieth century, all three groups had moved outside their customary domains, and the qualitative research typical of anthropologists and folklorists has enriched the sociological record.

At the same turning point, the rise of telectronics and their effect on globalization, and thus sense of place, have both renewed the angst about a waning sense of place (Archibald 2004, 1) and created a deeper understanding of sense of place. Norwegian rural sociologist Agnete Wiborg writes, "Instead of arguing that identity in modern society is becoming ever more detached from specific places, it may be more fruitful to consider the relationship between people and places as a changing and complex phenomenon, involving a shifting perception and meaning of locality and place, something which has implications for understanding belonging and migration" (2004, 416). Considering the lives of transmigrants, Debra Lattanzi Shutika concludes, "The restructuring of spatial relations through globalization certainly poses particular challenges to places, but it neither destroys them nor lessens their significance" (2011, 14). Quite the opposite, the transmigrant communities she studies lovingly tend to two sets of homeplaces in two different countries.

Writing well before these ethnographers, indeed well before cell phones and the internet became common let alone integrated, philosopher Edward Casey (1993) takes on sense of place as a broader conceptual issue. He sees a shift, taking centuries, in which *time* replaced *space* as the primary organizer of human experience. He notes, "We calculate, and move at rapid speeds, in time and space. But we do not live in these abstract parameters; instead, we are displaced in them and by them" (Casey 1993, 38). In order to reclaim the power of place lost in this shift, we may start with the conclusions of Casey (1993); Wiborg (2004); Savage, Bagnall, and Longhurst (2005); and Lattanzi Shutika (2011): the death knell for sense of place has not yet rung. Rather, by considering the findings of ethnographers since Casey, we can use an increased, nuanced, and far more complex understanding of sense of place to make decisions that affect everyday life and worldview.

Readers may experience the elements that produce the complexity of sense of place as "common sense," for they are talked around in the media and in everyday life. However, popular discussion often floats on the surface and depends upon assumption. Connecting the elements of sense of place to different works of in-depth scholarship will enable us to get a fuller picture of the concept's complexity so that we may further analyze narrative stewardship's role in sense of place on Lewis Island.

Perhaps the first issue in the complex configuration of place is the combination of mobility and attachment, which in scholarly and lay conversation have usually been assumed to counteract one another. Wiborg notes that earlier researchers thought of people's relationships with places as being *either* staying there *or* going somewhere new (2004, 416). However, Wiborg argues, "There is now an increased awareness that mobility, and a more fragile and shifting relationship between place, society, and culture, in fact characterises [*sic*] many people's experience" (2004, 416–17). Scholars disagree about whether the increase in *mobility* increases or decreases attachment to *place.* "Attachment to places" has become "an object of negotiation and reflexivity" (Wiborg 2004, 417) and "is a multidimensional and dynamic phenomenon where different aspects of a place can be used, as practice, in the construction of images or as narratives in the formation and management of identity" (Wiborg 2004, 429). In Wiborg's study, this happened when Norwegian students in an urban setting reflected on their rural homeplaces. On Lewis Island, however, it can happen when visitors tell touchstone stories to Muriel Meserve, the island's co-owner: these visitors are negotiating their attachment to the place, both reflecting on their relationship to it in the past and establishing that relationship for the present and possibly the future.

Rather than simply seeing "places as entities," people can see them as much more complex phenomena, "includ[ing] varied and interrelated aspects of people's relationship to places, where place is differentially conceived as a geographical locality, a system of social relations and a way of life, or as a basis for symbols, a relationship that might include combinations of all of these" (Wiborg 2004, 417). Further, "construction of identity and ascribing meaning to places can be seen as mutual processes" (Wiborg 2004, 417; see also Lattanzi Shutika 2011, 15–16). This view resembles Casey's view on places: "You are *in* them not as a puppet stuffed in a box—as would be true on a strict container view of place—but as living in them, indeed *through* them. . . . They serve to implace you, to anchor and orient you, finally becoming an integral part of your identity" (1993, 22). Putting this last idea together with the well-established view that place is space mediated by humans (Tuan 1977), people's and places' identity and significance become contingent upon one another and thus ever evolving.

The complexity increases when one considers that the human-mediated, social, or cultural elements of place include both individual and collective contributions and interpretations. Kent Ryden describes it this way:

> Within the shared sense of place of any region there is, of course, a great deal of individual variation. Although all residents of a place participate in the same broad patterns of geographical and historical experience, each person's sense of place is altered to fit his or her personal experiences with those larger patterns: like jazz soloists, individuals create unique styles and interpretations of life while remaining within the rhythms and structures of a larger composition. (1993, 198)

Muriel Meserve reflects on the variation in people's identification with Lewis Island and what makes the place peaceful: "For me ... it's because that's home. That's where I grew up. So, I don't know what it is. I think ... it's a little bit different for everybody. But I think the river's a big part of it, but uh ... there's just something about it. I think it speaks to different people in different ways" (October 5, 2009). She goes on to discuss the island's spiritual significance to different people: "I think there's ... a little bit of a lot of things over there, so it's whatever you're looking for, you can pretty much find it" (October 5, 2009). Different people develop relationships with the island using what they bring with them ("what you're looking for"). Her comments echo the story her father used to tell about visitors seeing in a new place what they saw in the people where they were from (recounted in chapter 2), but she also draws on her experience talking to visitors over decades to demonstrate how one place can mean something different—and mean it in different ways—to different individuals. Moreover, as people discuss a place with one another—as they do with Muriel Meserve on the riverbank—individual versions are articulated and affirmed while the "larger composition" is co-created. Barbara Johnstone adds to our understanding of place meanings by reminding us of the relationship between public and private stories: "Private stories and individual ways of telling stories give rise to shared, public stories, and ... public stories form part of the context of private ones" (1990, 13). Again, the sense of place becomes exponentially complex as individual stories about place are shared among individuals, then repeated to other audiences and revised in retelling to eventually create shared (and sometimes contested) public stories and a more-or-less shared sense of place.

Savage, Bagnall, and Longhurst's concept of "elective belonging" (2005) becomes very important here, for it well describes the varied and simultaneous, and thus also multivalent, influence of individuals on sense of place. Using interview to study people's sense of their geographic communities around Manchester, England, Savage, Bagnall, and Longhurst make a significant shift in viewing both belonging and sense of place. Rather than concentrating on the idea of whether people belong to a place, the authors emphasize choice: whether and how the place belongs to the people. By giving people agency over their own belonging, these scholars move away from the customary view that lifelong locals own a place and determine who else belongs; this shift makes sense of the fact that, as Lattanzi Shutika notes, lifelong locals do not always feel *they* belong to rapidly changing places themselves (2011, 11). People belong not because others accept them but because they choose to belong and dedicate themselves to a place. This choice, however, is not arbitrary:

> Individuals attach their own biographies to their "chosen" residential location, so that they tell stories that indicate how their arrival and subsequent settlement is appropriate to their sense of themselves. People who come to live in an area with no prior ties to it, but who can link their residence to their biographical life history, are able to see themselves as belonging to the area. (Savage, Bagnall, and Longhurst 2005, 29)

Thus, people belong to Lewis Island because they commit to it—or do not belong because they choose not to embrace the place. Commitment might appear as dedicating one's time to being on Lewis Island, perhaps fishing, buying fish, walking the dog, or visiting, or to interacting with others on the island, often taking part in narrative stewardship. People's motivation for commitment will depend on some combination of practical situation and individual characteristics, such as Ted Kroemmelbein's conviction that he was "born ready for it" (June 15, 2013). With the concept of "elective belonging," Savage, Bagnall, and Longhurst not only develop ideas of belonging but also develop the concept of sense of place—what people belong *to* and what belongs *to them*. People are agents and not just subjects of belonging in a process that has wider impact, because they can define places as they define themselves, themselves in relation to places, and places in relation to each other. Considering how individual will affects belonging and perception entails an important shift in understanding the construction of sense of place: places are unstable, are interdependent with identity formation, and involve a complex interaction of various individual and collective influences.

Individuals in combination then form the sinews of communities and social networks associated with places. As commentators have predicted the end of places, so too have they predicted the end of communities and face-to-face communication (Bruhn 2005, 102–3; Wiborg 2004, 429). These ends are usually attributed to factors such as "technology, rapid social change, the ineffectiveness of social institutions in meeting new needs, greed and selfishness, greater ethnic diversity, the loss of community as a 'place,' generational differences, changing values, and fear" (Bruhn 2005, vii). Not surprisingly, researchers argue to varying degrees that—like sense of place—community and social connection may change with challenges but will not be obliterated, particularly if people draw upon the power of narrative.

Barbara Johnstone, for example, notes that shared narratives helped preserve Fort Wayne, Indiana, when floodwaters threatened the place and its community, using what she calls a "community story": "a story that belongs to a group rather than to a person. Insofar as the community to which the story belongs is defined by a place . . . the story belongs to the place, and insofar as communities of people are constituted by shared ways of giving meaning to experience in stories, the study of stories like Fort Wayne's flood story is a study of the narrative texture of the community itself" (1990, 119). Newspaperwoman Renée Kiriluk-Hill speaks similarly of the Lewis Fishery, connecting stories, people, community, and place: "What is the Lewis Fishery about? *Connections*. Community connections, connections with the environment. . . . And what is Lambertville about? You know, Lambertville is very much what was, not just what is. And Lambertville *honors* what was, so the Lewis Fishery is a *big story* . . . in this area, because of that" (June 3, 2011). As though she were reading Johnstone's theory, Kiriluk-Hill continues, moving between casual and journalistic concepts of story: "A place *is* its stories. Like there's the saying that you can get now on plaques: home is where your story begins" (June 3, 2011). Johnstone neatly connects narrative, place, belonging, and community this way: "A person is at home in a place when the place evokes stories, and, conversely, stories can serve to create places. In a sense, a community of speakers is a group of people who share previous stories, or conventions for making stories, and who jointly tell new stories" (1990, 5). Savage, Bagnall, and Longhurst (2005) might then add that one's story—that is, one's biography and what one chooses to emphasize and connect with a place—creates place-based belonging. So, not only are individuals and places mutually defined but so are communities of people and places; despite increased mobility and more non-face-to-face interactions, attachment to place and one another can still exist through individual and collective agency, and through narrative stewardship.

Although Johnstone uses a conventional meaning of "community" that can be switched for "town" or "neighborhood," readers may think of the variety of communities on Lewis Island (e.g., townspeople past and present, school parents, crew members, dog walkers, people who fish, customers) who make up a more or less interrelated group of people who value the island, although not necessarily the same aspects or for the same reasons. With the tentativeness that characterizes youthful speech and postmodern experience, Keziah describes a shared value of community on the island in terms of "taking care":

> Well I think we gotta take care of a community, kind of. Well, there's like different people [older, former crew members] that come on [the island] . . . you make sure that they're doing okay and stuff like that . . . and I think also to each other, like we were having trouble when *Dad* [was ill], and Pam brought over a lasagna, things like that. And you just kinda make sure that everyone's doing okay. I feel like, a lot, when we're there, the whole crew kind of takes care of the kids too more. It's *harder* for the kids, because you have so many people bossing you

around and it gets really hard but it's like, you just have to think it's like the whole village raising a kid, more or less, yeah ... (October 3, 2009)

Here, she sees a give and take in maintaining the community with positive and negative aspects, but overall, an ethic of place-based caring emerges to which she applies the contemporary proverb that gained popularity in her childhood: "It takes a village to raise a child." She also notes that caring relationships connect the island and the wider community. Muriel Meserve asserts that "there's *always* been that sense that it's [the island is] a part of the community, it's there for the community to use, and it's a place that people choose to ... mark events in their lives" (October 5, 2009). *Choosing* to mark events (such as weddings, memorial services, and proposals) on Lewis Island reflects the fact that people "elect" to "belong" to the island, and it to them, activities encouraged by the extended Lewis family because of their mandate to share the island.

While sense of place begins with the physical (Tuan 1977, 183–84), the mandate to share the island is one aspect of the other part of sense of place: culture. Edward Casey (1993) explains the richness of the particular combination of landscape and human influence that creates sense of place, which he calls "thickening":

> The conjoining of nature and culture is not just a combination, much less a compromise or a mere synthesis. Something is gained, something *emerges* from the conjunction itself that is not present when the factors are held apart. Let us call the process of this emergence "thickening." By this I mean the dense coalescence of cultural practices and natural givens. In undergoing a mutual thickening, these practices and givens sediment and interfuse. (Casey 1993, 252–53)

Casey's description of a *process* underscores the point made above that sense of place is contingent, multivalent, and ever changing, even while one can experience and describe sense of place through evidence or imagining that thickening has taken place. In his treatment of place, Casey uses Thoreau's term "cohabitancy," which Casey defines as "a special kind of settled coexistence between humans and the land, between the natural and the cultural, and between one's contemporaries and one's ancestors" (1993, 291). "Cohabitancy" well reflects the fishery family's experience of accepting the river's peace *and* danger, carrying out the mandate to share the island, and living the other Big Stories. When "ancestors" are understood more broadly to indicate various people who have come before one, others experience "cohabitancy" through their experience of the island's landscape and culture, particularly through touchstone narratives.

Our discussion of the anatomy of Lewis Island's sense of place should include, as Yi-Fu Tuan (1977) suggests, the physical—but not before we are reminded that the physical is not static, either. Various Lewis family members, crew members, and visitors whose experience spans decades comment on changes to the island. Ted and Diane Kroemmelbein, for example, comment not only on the seasonal changes that occur with each day's passing but also on the toll that flooding takes on the island and river channel, the cycles of plant and tree growth, and the healthy changes to the river related to pollution remediation in the mid- to late twentieth century (June 15, 2013). Ted Kroemmelbein calls forth some of the same features but emphasizes looking at them carefully because they change:

> I love looking at the water to see what's in there and how it is, and how clear it is today. And at the end of the day, I'll always sit on the [bench on the riverbank].... I love watching the sun. You know ... wherever I'm at this time of day, I'm just outside watching that. That's what I like to do, and then the island's a special place for that. (June 15, 2013)

As with elective belonging, Ted connects something in his biography (stopping to enjoy the sunset *every* night) to this place, noting both its specialness and its place in his life. Mary Iuvone, a photographer for a regional newspaper, likewise unites the sunshine and the work in her memory of covering the fishery: "The first time I remember, because it was just such a beautiful night. It was sunny. It was sort of warm, and I was there until sunset. And it was really beautiful, you know, the whole process of what you guys do, the teamwork" (May 20, 2011). For Mary, too, the island at sunset is inherently connected to the fishing activity. Ted's reference to sunset also relates to the fact that as a Jersey fishery, the Lewis Fishery is an "evening fishery," with the shad flocking to the warmth of the eastern riverbank in the late afternoon and into evening. Reflecting on her lifelong relationship with the fishery, Diane Kroemmelbein also connects the light with the fish in the river: "That's the thing that I always loved to see was the glittering of the silver and the gold, when the golden carp were in there, and the shad were in there.... The sun would hit on it, and it would just be one big glittering mass" (June 15, 2013). Diane's statement refers not just to a river but to *this* part of *this* river. In the speech preceding this statement, her husband had just noted that the river changes with flooding, and so only in particular years is the river channel shaped such that the fishing crew can reach and catch the carp using the haul seine method. Thus, the natural landscape and the cultural foraging activity of haul seine fishing coalesce in this particular sense of place.

In addition to light and water, the plant life creates a changing yet constant sense of place on Lewis Island, where particular plants are expected in particular patches of earth at particular points in the spring season. In the absence of heavier foot traffic over the winter, the island plants reassert themselves and green up the island again. Ted Kroemmelbein points to the tension between human influence and natural growth when he describes the island: "What I like about the island is that, except for a utility pole, some electricity there, there's nothing else there, you know: it's empty. It's pristine as it can be, you know . . . It's the same but it always changes" through the natural season from snow to heat, with different flowers blooming over the course of the fishing season (June 15, 2013). Ted communicates an appreciation for the known and the unknown; he uses the names of some plants, but others are just "little yellow things" and "little white things." He knows the general trajectory of the plants' season, but each day is still a surprise to him when he experiences exactly what it is like in this place. Here, the landscape shapes the human experience, even as the humans shape the landscape through the built environment, human activity, and human perception.

However, the more enduring fact that the island is an island, defined by the river and its separation from the mainland, should not be underestimated as a key feature of the place. Dan Tuft explains what makes the island special in just these terms:

> I think it's being right by the water: you feel like you're almost in the river, and you get the great view of both towns on either side without feeling like you're right in the middle of it, so you're kind of stepping outside of town at the same time that you're very close to it. And . . . I notice the sky a lot more when I'm down there. It's just very peaceful. Even though it's been flooded, it still feels very peaceful. (October 3, 2009)

Dan emphasizes the island's relationship to the river, towns, and sky in a liminal place between all three. Moreover, the island's position becomes the position of those on the island, creating a unique experience.

Again and again, as people describe the island, many mention "peacefulness" while others use synonyms such as "quieting" (Steve Meserve, October 3, 2009) or "calm" (Pam Baker, October 4, 2009). Pam Baker and Dan Tuft in particular cite the "peacefulness" or "calmness" as benefiting children and teens (Dan Tuft, October 3, 2009; Pam Baker, October 4, 2009). Just as Dan does, Muriel Meserve credits the landscape features of the island and river with creating this peaceful sense of place:

There's just a *feeling* once you're on the other side of the bridge, that you've left all the rest of the world behind you. I don't know, it's a sense of peace over there that you don't get anywhere else. I can't tell you how many people have said that they've worked out their problems out there on the island.... And I know that always, that's always been ... my place to go, you know, if you really just want to sit and think. That's the place to go, because there's nothing like watching that river. It kind of puts everything into perspective for you. Then you can come back across the bridge, come back to the real world, and it's okay [laughs]. (October 5, 2009)

In another interview, Muriel describes these people as having "a religious experience," "having it all come together and understanding what to do, and being totally calmed down by being there.... The island has that effect on people" (November 7, 2003). Over the years, Muriel's father, Fred Lewis, referred to something similar, calling the visitors' activity as "walking out their troubles." His grandson, Steve Meserve, cites the same phenomenon:

I don't know if it's because of the river there and the sound of it, or just its presence. The obvious history of it, you know, through the millennia, or whatever, but it just seems to be a very quieting place, and it always has been. My grandfather always told stories about people who would come over there with problems and be able to leave with them solved. And all they would do is just sit by the river and think.... It's just enough knowing that that's possible there, I think, for a lot of people, that to try to figure out the whys is pretty useless. (October 3, 2009)

Steve's reference to the "millennia" seems to call forth the idea of cohabitancy again: humans live a life that is integrated into the landscape, which connects them with other generations of humans. The same summary of troubles-resolution events from three different people suggests that such occurrences are a layering of narrative upon the island to create sense of place—the "texture" that Barbara Johnstone refers to: "Coming to know a place means becoming a character in its stories and making it a character in yours, and memories of places are often organized around memories of events in which one had a part" (1990, 10). Both the visitors and Fred Lewis become characters, and Muriel and Steve "had a part" in these events as either the person who solved a personal problem, the person who interacted with such a visitor, or the person who listened to such a story told by Fred Lewis.

Natural settings are texturized not only by the people themselves but also by material culture. Edward Casey argues that buildings are obvious cultural influences, but he also nuances the concept by attributing to buildings a liminal quality: "[A building] exists between the bodies of those who inhabit or use it and the landscape arranged around it" (1993, 32). On Lewis Island, the construction of house, shed, crew's cabin, and bridge reflect minimal development. These structures, a liminal space itself, exist even more liminally, for the river repeatedly damages the buildings and bridge, and the people repeatedly rebuild them with the river's nature in mind. In effect, these structures exist in a liminal temporal space between floods, having just recovered from one or standing in anticipation of the next, as well as between water and earth (island) places.

Other human-made structures on the island lie between buildings and landscape on the spectrum between nature and culture. The park-like benches reflect the communal desire to accommodate several visitors enjoying the river at a time. The stone paths and spots for bagging up during the haul made by Theodore Lewis (Uncle Dory, Bill Lewis Sr.'s brother), mold the landscape to the activity of haul seine fishing, but the river has also at times covered over these places with silt. Although people were upset by the destruction of the floods in the mid-2000s, old-timers had to admit that the erosion also served to uncover some of Uncle Dory's work, thus restoring the island to an earlier state with more visible enculturation. On a daily basis, the river covers or reveals particular stones on the side of the island as the river height fluctuates, and the captain reads these signs—a cultural element—to determine how to fish that day. As the extended Lewis family and friends respond to flooding, they may rebuild the human-made structures but intentionally leave the technology and materials simple (Sue and Steve Meserve, October 3, 2009). Meanwhile, many human-made objects—refuse from upstream—are removed from the island.

The island's woods and natural groundcover undergo other types of enculturation. At one time a horse was used for continual growth control, but in recent years Weedwackers and lawnmowers have replaced it. In a retro move in the summer of 2014, however, a few borrowed goats were applied to the poison ivy. The plants along the center path crowd pedestrians as the growing season progresses, and the path must be periodically reestablished with a mowing. The center path leads up to Mrs. Berg's garden, which by its horticultural nature is plants-plus-culture, in particular the Swiss culture of Mrs. Berg's youth. Moreover, when Fred and Nell Lewis allowed Mrs. Berg to establish her garden on their property, they follow their family practice of sharing within the larger and contradictory capitalistic culture, enculturating Lewis Island differently and influencing its sense of place. In a final move of enculturation, some of the children who visit the island use the name "the secret garden" interchangeably with "Mrs. Berg's garden," making a literary reference to Frances Hodgson Burnett's century-old but still popular children's book, *The Secret Garden* ([1911] 1962), in which three children, differing in temperament and social station, work to restore a garden, building physical, moral, and emotional strength. Pam Baker refers to the many activities kids can do with the simple and natural resources on Lewis Island when she says, "You know, it's all right there . . . we don't have to create anything. We don't have to have the ping-pong table and the this and the that; it kind of takes care of itself" (October 4, 2009). The island place can make the party because its natural resources tap into traditional cultural pastimes like walking, swimming, boating, no-tech games, and simply sitting and "connecting," as Pam says (October 4, 2009).

The place is also inextricably connected to the haul seine tradition, and the activity would not continue as it has anywhere else for both geographic and cultural reasons. Obviously, the activity owes its existence to the fact that shad return to their natal waters to spawn. The family frequently recalls an article published decades ago in a regional paper that found the best-tasting shad between Scudder's Falls and Bull Island, and Lambertville sits about halfway between those two Delaware River locations. The shad pass Lewis Island in the middle of their run, when the fish have become less oily, having lived off their own fats, but the eggs have not yet lost much of their flavor, becoming more blonde and translucent only later in the run. Therefore, the Lewis Fishery has the most delicious of the "most delicious shad" (which is the literal meaning of the species' Latin name, *Alosa sapidissima*), a culturally informed aesthetic anchored to a natural location. As an evening fishery, the Lewis Fishery aligns the most naturally successful time for fishing with a culturally structured time: the crew can work at the fishery after the conventional close of business. Further along the cultural end of the spectrum, the fishery's fishing license is rooted to the specific spot, and, given that the family with the license also owns the island, they have more resources with which to keep the enterprise going.

Although much of the island's relationship to fishing seems natural, the shift to the human activity is inherently cultural, and other cultural practices spring up through engagement with the space. The process of adding people and culture to nature, which Casey calls "implacement" (1993, 31), includes the mapping of each part of Lewis Island used during the haul. Moving on the outside path (a place), the captain will call out to the crew where to stop, using names that have themselves become traditions, such as "by the buttonballs" (sycamore trees). Out in the river, the crew will measure their position by counting the pylons and uprights on the bridge just downstream as well as by fractions of the river's width (three-quarters, one-half) to decide where, when, and whether to change the pace of rowing, and when to "turn toward home," the island being defined as "home" regardless of where the crew members reside. The members of the crew on the bank will have asked the captain where he expects the boat to come in: by Jake's dock (now just a small outcropping created by overgrowth covering the location of the old dock and commemorating a relationship with someone else the family shared the island with), by the stairs (also overgrown, but functional), by the house, or by the cabin. Bagging up will then happen "on the side" or "at the point," and that geographic decision will have determined whether someone will have "stayed down" to rake or sweep the point and talk to visitors, interfusing a fishing role, a social role, a narrative role, and a place.

A far more specialized sense of place is that of the captains, rowers, and landsmen, who come to understand the changing river through its currents' effect on the boat, the net, and the fish. Steve describes a narrative stewardship process specifically in relation to traditional river knowledge:

> We're always, *I'm* always thinking back to, because nothing's written down. It's all oral history, or *oral* manual as far as the operator's manual for this thing that we do. I have to go back where [laughs] things were when I fished with Grampy, and what we did, and his telling of how to do it at these certain times, when the water's at certain levels, and, well, here are the marks to look at, where is it on the concrete, where is it on the bridge, where is it on the shore. So, there's no graph or chart or anything to go by. It's all kind of by feel and through the experience of the conversations that we had as I was growing up. So as far as that piece, of figuring out what size net to use, what length, those stories come back to me—and even now, at an age where I can also go on my own experience—[are] things I talk about, and talk to others about, so that I try to pass along some of that knowledge as best I can. (October 3, 2009)

Repeating the same salient features of reading the river that Fred relayed to me over a decade earlier, Steve describes the captain's map of the place. What's more, he sees himself first as audience and then as narrator, in the process of stewarding river knowledge as a communal resource.

Speaking from a landsman's point of view, Ted Kroemmelbein says that when he learned that role, he felt he needed to ignore what the previous landsmen told him, which he came to believe was not true or useful but rather just "what they'd been told, and it [doesn't] necessarily mean that that's, you know, the Bible" (June 15, 2013). Ted now understands the relative positions of the net, himself, the river current, the boat, and the river's edge in relation to the positions and workload of the crew later in the process—in essence, the arc of the emerging haul:

When I first started doing [the landsman job], I would work like crazy, because they said, well, when you get here, you've gotta have it like this at this angle. Well, you *don't*. If experience tells me if you just pull back a little bit, gravity will take care of [straightening the net] in about 150 more feet, and you're already there. . . . So, your object [in doing] land is to make it as easy as possible with the least amount of work for *everybody*, so you control what's in the boat, how the net comes around, how much, how hard everybody has to pull when they're down the end of it, depending upon how close you've got the net. When it all comes around and starts coming together, are you way out there where you don't need to be? Or are you in further, and—you ever notice there's no more yelling anymore about "hold back, Ted, hold back, Ted," or "let go, Ted"? There's none of that. I got that down to a science. You know, and . . . it makes it easier for everybody. So, that's my job. (June 15, 2013)

Ted's description shows that his role (an inherently cultural concept) is rooted in nature but is influenced by the group process. Following an ethics of kindness, he builds and uses his knowledge of the place and the forces active in it to make less work for his team. He affects the social situation by reducing work and the friction of frantic yelling, thus upholding the peaceful sense of place on Lewis Island.

Photographers have a whole different map of the space and process, one devoid of traditional names but connected to their occupation and art, a basic understanding of the crew's work and the local history of how the fishery has been covered in the past. Kiriluk-Hill calls the place "*such* a photographer's dream!" (June 3, 2011) because of the light conditions at the time of day when the crew fishes. However, the space also poses challenges. Photographer Mary Iuvone maps the place according to the parameters of her work: "It is a little bit of an obstacle, because you have the water there. And so there's how many perspectives? I can be on the land or in the boat. You know, I would like to actually be in another boat and photograph from that boat, you know, you guys fishing. But that would involve getting someone to put me on another boat" (May 20, 2011). Moreover, she is concerned with the recorded visual history of the place: "Everybody's photographed it, so we're always looking at their pictures trying to figure out, what could I do different next time, you know, just to vary it up a little bit" (May 20, 2011). Iuvone's view of covering Lewis Island reflects layers of activity and media history and is thus inherently enculturated—in Casey's terminology, "implaced." As subjects, the crew members have also begun to notice a set number of representations of the place, referring to strategies they have seen photographers use over the years such as the shot through the net while loading or mending net, the shot of bagging up from the point looking out at the bridge to New Hope, and the shot of bagging up from the water looking up at the island. The crew members' conversations refer to images they see of themselves in print or on the web, when sense of place becomes mediated, public, and clearly enculturated.

The complexity of sense of place on Lewis Island becomes more evident as one considers the ideas of public and private place, with both misconceptions and changing ways of managing the space. Many people do not realize that Lewis Island is private property, which is particularly ironic, since many also think that the public land, below the bridge to New Hope where the Boat Club sits, is private property. This misunderstanding comes out in their expectations, such as when Ted Kroemmelbein reflects on "irate" Shad Fest visitors: "We'll get people going on the island that get very upset that there's no public bathrooms" (June 15, 2013). In a 2012 survey, I asked festivalgoers whether they thought the island was public parkland, and a quarter of the respondents answered that it was, while another quarter left the question blank or answered with a question mark.

Dan Tuft surmises that more people now understand the island's private property status than before, ever since the family made a larger "Private Property" sign in the mid-2000s to replace a smaller sign first erected in the 1970s. The signs reflect a tension between the mandate to share the island and the realities of risk management and concern for keeping the public safe when on the island. A lockable gate was built to block access to the footbridge during the frequent flooding in the mid-2000s, the same time period when leadership transitioned from Fred Lewis to the next two generations sharing the responsibility after his passing. Muriel Meserve and her children, Steve and Pam, felt very conflicted about using the larger sign and gate, which are the antithesis of welcoming people to the island they aim to share. However, they also recognized the fact that they could not monitor the island as closely as Fred and Nell did when they lived on the island and/or stayed within sight of it almost round-the-clock on most days. Because of today's awareness of litigation, the younger generations may need to be more risk averse than Fred was when it came to the possibility of accidents. Nevertheless the gate gets locked only when flooding or some other special situation poses danger to the public.

Ironically, in some ways the island was shared even more after the changing of the guard, sign notwithstanding. At this point, when the gates are unlocked, just the buildings and fenced yard around the house remain private. Back in Fred Lewis's day, however, Kelly McMichael remembers barriers put up to protect private family space when Fred and Nell allowed the town to set off fireworks from the island on the Fourth of July. The point was roped off for the people lighting the fireworks, and then, Kelly recalls, "Fred would rope off a part, like, we could go on this side and everyone else on that side (June 15, 2013). The "we" was the family that owned the island and their personal guests, who would have a barbecue before the event, and Kelly remembered that being included in the special section made her feel not only special but "like family." She recalls that her side of the rope was a much better deal than where the general public watched the fireworks: "I remember them cutting back trees to get ready for it so we would have like perfect view, and we were allowed on this section, and everybody else, you're on your own over there." She continues, "People were squooshed, and we had all this space," and describes the middle place between the family seats and the fireworks launching area as "like a free-for-all. Like if you could fit, if you could jam yourself in there, go ahead" (June 15, 2013). The description she gives sounds exclusive, to be sure, but one should recall the context: the island would have been the only private property open to the general public that night.

Eventually the fireworks were discontinued, but they were reestablished as a weekly event in the summers of 2011 through 2013 and on a reduced schedule thereafter. Despite the private property sign, the new generations stewarding the island have prepared even more to accommodate the public on the island. They strung a set of lights on the footbridge and took turns stationing themselves at the entrance to the bridge to welcome people onto the island. Once on the island, visitors are accommodated with a fleet of plastic chairs. Although the family might reserve a couple of chairs to ensure that their own elder members have a place to sit, the roping of Kelly's memory was gone. Rather, the family totes chairs to the point for strangers as well as family.

In one interview, when Muriel Meserve and I were discussing the fact that people don't tend to sit on their front porches anymore, she said that the island was "the town's biggest front porch, and *anybody* can come on it, you know" (October 5, 2009). The metaphor works well for the fireworks events, and it can be stretched to refer to the way the island has served as the town's backyard and playground as well for generations. To be sure, the place now includes a physical palimpsest of artifacts of long-ago social activities. Sue Meserve recalls:

> The stuff you find there, like when the floods took away all the topsoil and we found all those glass bottles, all the time, all the soda pop bottles. Come to find out that [Steve's] mother and her brother ran a soda pop stand under the cabin for years, and nobody collected the bottles, or what bottles did get collected got left someplace, and so that's why we found [them]. (April 22, 1996)[1]

Muriel also recalls the miniature golf course from the Great Depression years described in chapter 2. As happened to so many features on Lewis Island, Muriel says, "eventually it got flooded." She continues, "You know, but when we were kids, we would find golf balls out there every so often and Dad would say, well, that's from the golf course" (October 5, 2009). This is another episode in the island's enculturation as a place for fun, and while Muriel learns about it from stories, she also sees the evidence years later in the golf balls, now artifacts. In both Sue's and Muriel's memories, stories and physical items co-create "thickening," to use Casey's term, and physically create "texture" as well as the metaphor of storied "texture," which Johnstone (1990) posits as an element of place.

When telling touchstone stories, longtime and former townspeople indicate their understanding of the island as shared space and relate their experience of the place as saturated with memories, as implaced[2] and enculturated. Kent Ryden suggests that "experience, memory, and feeling combine with the physical environment to push peaks of human meaning above the abstract plane of space" (1993, 40). This heightened process of meaning-making resembles the way Sue Meserve moves between the soda bottles she finds on the island, the stories she has heard about people she never knew, stories of events that happened to her in-laws before she met them, and spiritual language to describe what makes the island special. Before giving her description of the bottles, Sue says to Steve,

> I just think it's all the spirits that live there [that make the island special]. I never met Poppy, but you feel like you know him, because you hear all the stories all the time, you know. I never met Eek [Fred Lewis's sister, Edith]. You feel like you know her. I knew your grandparents. I know, you know, your grandmother. I knew some of the crew that your grandfather [had], when it was Johnny and Robert and Terry. And, you know. I came in on that crew. And then there's other stories you hear.

Sue describes the artifacts that evidence some of those stories and then returns to the abstract and spiritual: "I mean, there's so much history, there's so many spirits, there's so many ghosts and all.... That's why, I feel, it's different." (October 4, 2009). Other crew members and family members have made similar comments about feeling the past all around them, in part through the stories' thickening and texture, which give the island profound meaning to them. On Lewis Island, the meanings emerge through the physical artifacts and space. When people move through the place and see the objects, they are concretely connected to the stories, and thus to the other times and people the stories are about.

Keith Basso describes a similar process involving moral narratives among the Cibecue Apaches on their reservation land in his classic article, "'Stalking with Stories': Names, Places, and Moral Narratives among the Western Apache" (1984). Just as the place names evoke the moral narratives that instruct the Cibecue how to behave, the objects and events on Lewis Island evoke stories connected to people who are themselves reminders of the Big Stories that communicate values and expected behaviors. The connection between civility and place has been clearly passed to the crew's young people. When asked about what makes the island special, Keziah spoke immediately about the higher behavioral expectations she had to meet there. She summarizes these expectations as, "You don't act like a brat," but the requirements go right down to facial expressions: "You can't roll your eyes or anything, 'cause you just can't do that," for it would "mess it up . . . 'cause that's like what I was saying before: the island is like—nobody's like that" (October 3, 2009). Here, she links the behavioral expectation to the island (eye-rolling might happen elsewhere) and implies that not following the norms would negatively change the place. Similarly, Sarah Baker connects the island to behavioral expectations:

> You know, on the island, that's one of the things I *like* is we're *constantly* nice to each other. There are *moments* that we get, you know, a little *tension* and there's a little anger, but you know we try to resolve it as quick as possible. We don't hate on each other. We don't hate on the people that's, you know, coming to the island. That's the whole *nature* and the *peacefulness* of it. (September 13, 2012)

Like Keziah, Sarah acknowledges that community members are not goody-goodies who always achieve the highest standards, but they do make an effort to achieve them more consistently *on the island* than in other places, and so those ideals become part of the "*nature*" (emphasis in the original) of the island, what scholars call the "sense of place."

Basso's example of noting the cultural import of place names reminds one that "Lewis Island" is itself testimony to the importance of the Lewis family and their Big Stories. Originally and legally, the island is Holcombe Island, but during the twentieth century it came to be known equally, even predominantly, as Lewis Island. Kiriluk-Hill notes, "People *call* it Lewis Island, but it's *not* Lewis Island. It's really Holcombe Island [laughs], and, how much more can you get that connection than people call it what it's not? [laughs]" (June 5, 2011). The Lewis family members use both names. Steve and Sue Meserve recall that the US Post Office recognizes the name "Lewis Island" (April 22, 1996), and the house on the island's address is "1 Lewis Island." By the 1996 interview, the film *Miracle on 34th Street* (Seaton 1947) had long been a well-known classic, and a remake had come out two years earlier (Mayfield 1994). At the climax of the film, the post office delivers to a particular man children's letters addressed to Santa Claus, and their decision is used to legally establish the man as Santa Claus, thereby evidencing his sanity and saving him from involuntary commitment to a sanatorium. In like manner, Sue and Steve's recollection of the post office as an authority adds culture to Lewis Island, not only through government authority but also as an artistic reference that puts symbolic truth above the historical record in an establishment of identity.

Both the Cibecue and people connected with Lewis Island not only live their moral lives in relationship to place but also build their identities as individuals and members of groups through connection with the place in a process akin to elective belonging. Kent Ryden aptly expresses the basic relationship between place and identity through memory: "If we feel that our present selves are inextricably bound to our pasts—that our lives have historical continuity, that we are the products of our past experiences—and if we tie memory to the landscape, then in contemplating place we contemplate ourselves" (1993, 39–40).[3] When I surveyed festivalgoers in 2012, I had a check-box for Lambertvillians to identify themselves, but some other people used the comment fields to identify themselves as people who used to live in Lambertville or were friends of Lambertvillians, and to attribute their knowledge of the fishery to geographic and social connections. Current and former Lambertvillians identify themselves with the geographic label when they visit the island, too, either by telling people on the island whom they don't know about their status as current or former Lambertvillians or by sharing a memory of Lambertville with another Lambertvillian who shared that experience. Barbara Johnstone analyzes the efficacy of such stories:

> Community stories serve the same sort of function as geography talk, in a less explicit way.... Collective knowledge of places evokes the collective memory that defines a group. Individuals' relationships to groups are mediated through shared memories, memories organized around places and the stories that belong to places. (1990, 121)

Where Ryden discusses self-*contemplation* in the quotation above, Johnstone emphasizes the *creative* process. She focuses on belonging and community: through the same storytelling process, people create their individual identities as well as co-creating sense of place through narrative. This happens whether the narrators live in that place or simply tell stories within or about that place, which can become symbolic.

Agnete Wiborg addresses the power of symbol when she concludes from her study of urban university students that "living in a place is not a necessary condition for attachment to it. The way the students express their attachment to their homeplace is dominated by the symbolic and emotional aspects of the place rather than the place as a material arena for social interactions" (2004, 429). In addition to Lewis Island and its fishery symbolizing a certain type of ethics and a history, through everyday storying visitors create identity *both* through the on-site social interactions Wiborg says are unnecessary *and* through symbol and emotion. For Wiborg's subjects, the signs and symbols relate to a variety of concepts, including "kinship, nature, rural lifestyle, class, gender, and so on" (2004, 429), and so it is for Lewis Island visitors. Through everyday storying, touchstones, and other utterances, visitors present any of the following: their kinship with the Lewis family, crew, or former visitors; their attachment to the river and island as a natural place; their experiences with fishing; their connection to a "simple" or "old-school" life they associate with Lambertville or the fishery; or gender identity factors relating to male fishing, female narration, or the gender subgroup of women who do traditionally masculine activities. These references may or may not relate directly to Lambertville or the fishery and may in fact connect Lewis Island to other places through a "symbolic geography" (Savage, Bagnall, and Longhurst 2005, 78–79). This enactment of "elective belonging" (Savage, Bagnall, and Longhurst 2005) can be seen, for example, when crew member Ted Kroemmelbein explains the relationship between his personal history and his aptness for crew membership, even if he was not the avid, expert fisherman outsiders (especially avid, expert fishermen) expect to find on the crew: "I'm one of those guys that spent my youth in the creeks and fishing and catching frogs and snakes and jumping off bridges and tubing before tubing was popular, all that kind of stuff, so, it was a kind of a pretty good fit for me" (June 15, 2013). Other visitors will also connect themselves with the peaceful and spiritual aspects of the island, as they, also, connect their own natures and biographies to the island.

While Wiborg contrasts being in a place and symbolic attention (see above), one should keep in mind that, in ritual behavior, doing something in a particular place becomes symbol. This is the "touch" part of the touchstone stories and the "pilgrimage" aspect of seeing shad haul seining as a rite of spring, as Muriel Meserve (November 7, 2003), Steve Meserve (October 3, 2009), Pam Baker (October 4, 2009), and Mary Iuvone (May 20, 2011), as well as many unidentified visitors, call it. In a neat conflation of nature and culture, Pam Baker observes, island visitors make visiting Lewis Island "part of their yearly ritual . . . just like the shad" (October 3, 2009). Being there then translates into personal history and tactile memory to create elective belonging, and when that experience is referred to in narrative—itself a ritual behavior that happens on Lewis Island—texture and thickening emerge as sense of place, and individual and group identities, are formed. Other forms of implaced symbolic action include processional storytelling and are found in the work itself. One responds to changes in the river and island as one responds to weather conditions that affect the haul, debris brought downstream that creates obstacles, and plant growth. Pam Baker attributes the fact that her grandfather allowed her to row during the haul with him on a particular Good Friday to her "sweat equity," a physical and sited investment in physically maintaining the island (October 4, 2009). As discussed above, the island is also a locus for enacting social behavioral ideas. In enacting the ideal, traditional, caring, face-to-face community (what Keziah calls "the *old* Lambertville"; October 3, 2009), symbol is wedded to experience as people create that community, even if only for a limited time in a limited space.

Kent Ryden emphasizes the importance of the power of place in creating identity when he writes, "This sense of identity may be one of the strongest of the feelings with which we regard places: when our meaningful places are threatened, we feel threatened as well" (1993, 40). This idea explains the mood when the island was closed for so long after severe flooding in the mid-2000s. The erosion, dangerous debris, water and mud in the house, and loss of not just a few electrical wires but the lone utility pole went beyond "threatening" the place to actually damaging it, necessitating closing the island to the public temporarily. For the general public, this not only brought the real impact of the flood to people who did not live or work in the floodplains but also impeded access to a river they had come to depend on and that, as established above, had created some of their identity.

For the family, a big party for Muriel Meserve's seventieth birthday in 2009 marked a turning point in the restoration of the island as a community space, even though it had been opened much earlier: "When we had Mur's party, I think that was one of the best things. There were so many people who hadn't been there in so long because we closed it, and also just because of circumstance. You know, you just don't get down if you don't *know* if it's open, if we don't have a party" (Pam Baker, October 3, 2009). While the party was a private event, it was a large one that mingled many groups within the larger Lambertville community, present and past, because Muriel had spent most of her life in the town and was active in a large cross section of community happenings. Concurrently with this party, which linked to both the culture of the space and a birthday rite, the island was slowly coming back to life, literally, as the work of family and friends to repair the island was augmented by plant regrowth. Pam recalls in 2009 that while the human activity was returning, it would never be the same because of the practical changes that had unfolded since the oldest generation had stopped living on the island (October 3, 2009). The undercurrent in her recollection is the change to the family as well as the change to the island. At the time of this interview, not only had Fred Lewis been gone for five years and still greatly missed but Nell Lewis's health was declining rapidly, and she passed away just weeks after this conversation. Here, the family's identity is tightly entwined with the island; even though strength and happy times return, there are permanent losses and significant changes to both, and yet community, identity, and sense of place are sustained.

NINE

Fishing in the Mainstream: Anomie, Sustainability, and Narrative Stewardship[1]

> Cultural analysis is intrinsically incomplete. And, worse than that, the more deeply it goes, the more incomplete it is.
> —**Clifford Geertz,** "Thick Description: Towards an Interpretive Theory of Culture"

> We study others so their humanity will bring our own into awareness, so the future will be better than the past.
> —**Henry Glassie,** *Passing the Time in Ballymenone*

> UNLESS someone like you cares a whole awful lot, nothing is going to get better. It's not.
> —**Dr. Seuss,** *The Lorax*

When I first began studying the Lewis Fishery more than twenty years ago, it pretty quickly became obvious that Lewis Island was special, but it took the next couple of decades to figure out how. The conceptual journey took me through ideas of faith, narrative, place, time, and commitment, swinging between flashes of insight and flashes of *duh, that's obvious.* Anthropologist Clifford Geertz's reflections on thick description are both reassuring and deflating:

> Cultural analysis is intrinsically incomplete. And, worse than that, the more deeply it goes, the more incomplete it is. It is a strange science whose most telling assertions are its most tremulously based, in which to get somewhere with the matter at hand is to intensify the suspicion, both your own and that of others, that you are not quite getting it right. But that, along with plaguing subtle people with obtuse questions, is what being an ethnographer is like. (1973c, 29)

Coming back across this passage late in drafting this ethnography led me to think maybe I had "gotten it right": ironically, I was passing by the same thought-place as an anthropologist whose work I respect *by* not getting as far as I would like in my argument *and* doubting my findings *while* also suspecting that they are self-evident.

Eventually, I conclude that what's so special about Lewis Island is that it isn't special—or at least it doesn't have to be. Here, I attempt to follow Henry Glassie's advice in *Passing the Time in Ballymenone*: "Our study must push beyond things to meanings, and grope through meanings to values. Study must rise to perplex and stand to become part of our critical endeavor. We study others so their humanity will bring our own into awareness, so the future will be better than the past" (1982, xiv). On Lewis Island, place and narrative fit together in a complex and contingent, meaning-rich, organic structure of individuals, groups, identity, belonging, landscape, and culture ... so what? It is sufficient to simply understand the complexity of these interrelated concepts as lived in a traditional site, but we can also go further to see how narrative stewardship and the anatomy of sense of place and belonging can apply to the ongoing challenges that come with our historical moment. We can proceed in hopes that some findings may apply to other historical moments. Lewis Island is a "whole 'nother place," and yet it is not. There is more and less than meets the eye; over 125 years of sustaining meaningful relationships and activities in one place takes commitment and creativity, and with awareness, more creativity, and more commitment, can provide an example to help people sustain other meaningful aspects of their cultures in this time of rapid global and local change.

It may be time to consider possible intersections between particular cultural changes and a new wave of "anomie." This term, introduced by French sociologist Émile Durkheim at the end of the nineteenth century and studied by American sociologist Robert Merton in the 1930s, has been widely accepted as a basic concept in cultural study (*Encylopædia Britannica* 2016; Nanda and Warms 2010). Originally used by theologians to denote "lawlessness" (*Oxford English Dictionary*), "anomy" or "anomie" as conceived by sociologists and others is a state in which rapid social changes accompanying rapid industrialization outstrip the pace of cultural change, cultural change that could have helped people adjust to the social and practical changes around them. In a state of anomie, confusion about what to do, think, and feel can lead to rootlessness, lack of relevant mores, lawlessness, rudeness, hurt, anger, conflict, frustration, depression, violence, and suicide.

Today, a technology-induced social upheaval similar to that of the industrial age has resulted from the rapid and widespread introduction of technologies that are all the more influential on human relationships because they are primarily telectronic communication technologies. These changes have affected social interaction in a wide range of arenas from the home and family to school, business, government, media, and even war zones, and so conditions are ripe for anomie. Unfortunately, we are so often swept up in the mutually reinforcing American myths of ever-increasing technological, economic, and social progress promoted by large and powerful social institutions that it is difficult to step back, slow down, and apply what we have learned about industrialism and anomie to this new situation.

We might do well to draw on our experience with industrial-era anomie to avoid some of the negative by-products of rapid change by harnessing the power of narrative as well as the power of place, which philosopher Edward Casey calls "considerable" (1993, 21). Historian Robert R. Archibald writes in *The New Town Square: Museums and Communities in Transition* that "the most profound dilemma of this new century, inherited from the last, is a deepening crisis of place, and the accompanying ennui of placelessness" (2004, 1). His use of the word "ennui" recalls the aimlessness associated with anomie. Casey himself references anomie directly, claiming that it "often stems from atopia" or placelessness (1993, xi). Casey sees nostalgia as "not merely a matter of regret for lost times" but "also a pining for *lost places*, for places we have once been in yet can no longer reenter," and believes that "our own culture suffers from acute nostalgia" (1993, 37). Reading Casey's work a generation after he wrote it, one wants to slap oneself on the forehead and ask, if Casey was writing before cell phone usage and the internet became so widespread, why has our society not been able to use widely accepted nineteenth- and twentieth-century sociology as a clue to stave off twenty-first-century anomie? After all, it has now become almost cliché to observe that the particular nature of telectronic developments has affected our worldview by challenging our very concepts of place, time, and social relationships. Repeated expressions of this idea in the popular press often accompany positive or negative judgments, with giddy excitement or anxious dismay, respectively.

Meanwhile, sociologists and psychologists, operating the slow, methodical cogs of scientific study, have already begun to produce findings that link social media and electronic communication use to social and/or psychological vulnerability (for examples, see Alter 2017; Turkle 2015, 2011; Roberts and David 2016; Becker, Alzahabi, and Hopwood 2013; Thomée, Härenstam, and Hagberg 2011; Chesley 2005; Katz 2006; Landoll, La Greca, and Lai 2013; Bauman, Toomey, and Walker 2013). Unfortunately, the forces of commerce and technological development move faster than research in the social sciences and humanities, and so anomie activates and grows faster than related sociocultural knowledge and protective tools do. Moreover, social science, having an empirical basis, looks backward, while forward-looking humanistic scholarship gets less attention. Also unfortunately, the powerful nature of those social institutions (e.g., big business, the STEM fields, government, education) pushing technological development and adoption agendas ("It's the wave of the future!") has hampered our collective ability to think critically and to integrate the findings of social sciences and the humanities into our everyday experiences of technology in a capitalistic culture. Sherry Turkle's work as a psychologist attached to a technology-focused academic institution, the Massachusetts Institute of Technology, provides strong models for what I call "remembering to put on our critical thinking caps when considering technology," and what Turkle calls "realtechnik." Realtechnik, she explains,

> suggests that we step back and reassess when we hear triumphant or apocalyptic narratives about how to live with technology. Realtechnik is skeptical about linear progress. It encourages humility, a state of mind in which we are most open to facing problems and reconsidering decisions. It helps us acknowledge costs and recognize the things we hold inviolate. (2011, 294)

As a folklorist, I would add that examining traditional and everyday practices is an apt vehicle for riding the edge of continuity and change, for recognizing the benefits but also the costs of technological adoptions. This study of cultural material often taken for granted helps us to, as Turkle says, identify important values. Knowing what we value can enable us to live out other values in addition to devotion to technological development. Further, in cases where blind technological adoption endangers other valued practices, that awareness can empower us to modify or abandon damaging technological behaviors. In short, cultural knowledge helps us use technology rather than allowing technological industries to use us.

In her 2011 work *Alone Together: Why We Expect More from Technology and Less from Each Other*, Turkle uses original and published research to carefully examine changing human relationships with robots and telectronics, which in turn influence how people pay (or don't pay) attention to each other. She then traces the implications for concepts of time and space, and also for healthy social interaction and psychological health. Not rejecting technologies but rejecting others' shrugs of powerlessness over technological uses considered negative (for example, texting incessantly, even during funerals), Turkle says: "A shrug is appropriate for a stalemate.... It is too early to have reached such an impasse. Rather, I believe we have reached a point of inflection, where we can see the costs and start to take action" (2011, 295–96). In *Reclaiming Conversation: The Power of Talk in a Digital Age*, Turkle then focuses on the nature of conversation itself, writing: "My argument is not anti-technology. It's pro-conversation. It's time to put technology in its place and reclaim conversation. That journey begins with a better understanding of what conversation accomplishes and how technology can get it its way" (2015, 25). I certainly agree, based on my own observations in physical *and* electronic family and community life, physical and electronic workplaces and classrooms—and also based on study of narrative stewardship on Lewis Island. The theoretical tools that we can use to spot signs of anomie (e.g., heroin addiction and suicide) have long existed. Also available are theoretical tools related to cultural sustainability, belonging, and sense of place with which to pursue protective and healing measures against anomie.

In the mid-1970s, Yi-Fu Tuan contrasted our longings for place ("security") and for space ("freedom"), positing that "space and place are basic components of the lived world; we take them for granted. When we think about them, however, they may assume unexpected meanings and raise questions we have not thought to ask" (1977, 3). As cyberspace has become part of our worldview in ways unanticipated by its developers or still not fully understood, we might ask about our longing for the security of place as something we can know more fully and easily when the "freedom" of cyberspace has emerged and intersected with postmodern alienation. Casey recommends "re-implacement" rather than nostalgia (a recycling of time) (1993, 39). Recognizing human mobility as a norm (particularly in American society), Casey does not argue for staying put but instead integrates movement into the concept of re-implacement: "Getting back into place ... is an ongoing task that calls for continual journeying between and among places.... As travelers on such a voyage, we can resume the direction, and regain the depth, of our individual and collective life once again—and know it for the first time" (1993, 314). Casey suggests that we not try to reproduce the past but, rather, intentionally steward our relationships with place more in balance with our concern for time. In so doing, we can expect that our social relationships and identities will change positively in comparison to what we experience now. The problem is not how to preserve specific senses of place, community, and identity but how to sustain *for* places, communities, and individuals those parts of our worldview and practices that promote wellness and thus resist this second round of anomie.

In turn, this task challenges us to flip a proverb and actually "*see* the forest for the trees." Worldviews, and the practices that reflect and perpetuate worldviews, are embedded and invisible in everyday life, which itself is often disregarded as unimportant. Speaking of anthropologists' pull toward studying the "exotic," Geertz suggests that these subject choices are "essentially a device for displacing the dulling sense of familiarity with which the mysteriousness of our own ability to relate perceptively to one another is concealed from us" (1973c, 15). Here, he describes a phenomenon of which folklorists and social historians are only too painfully aware when we choose as our subjects our own cultural traditions and those of the ubiquitous cultural mainstream.

Before moving on to more fully explore the intertwined roles of sense of place and narrative stewardship in cultural sustainability, it may be helpful to summarize here how the different narrative types covered in chapters 2 through 6 relate to sense of place theory. The Big Stories, as myth-like narratives, all function as symbolic narratives and help create "thickening" (Casey 1993) and "texture" (Johnstone 1990). When they function as moral instruction, they also can build community well-being and satisfaction by providing "a shared sense of meaning and purpose" (Hawkes 2001, 13). Committing to further the values and aims of the Big Stories can then build not only attachment to place but also identity, belonging, and a sense of place based on personal or cultural values. Microlegends draw out wonder and contribute to a lively, textured, common local history, creating both belonging and sense of place. People become imaginatively engaged with emerging legends, helping them make sense of their own experiences with dangers such as flooding and vandalism. Everyday storying is the basic stuff of creating and combining networks, as well as individual impression management and thus identity development. Processional storytelling, finally, is the adhesive that attaches the people, activity, and narrative to landscape as well as to each other. At its base, narrative stewardship stewards the place and its natural resources, the culture, and the sense of community while also impacting the identity and implacement of the individuals who actively participate in the narrative system on Lewis Island.

At this point, let us turn to the concept of sustainability to further our understanding of how culture works on Lewis Island, stewarding resources organically, subtly embedded in vernacular traditions, at once largely invisible and obviously persistent, both marginalized and powerful. *Sustainability* became a key concept at the turn of the twenty-first century, starting first with environmentalists and moving quickly to applications in other areas, including cultural sustainability.

Writing for an Australian public planning audience but offering comments of use in other situations, cultural analyst Jon Hawkes breaks down the key features of sustainable ventures in *The Fourth Pillar of Sustainability: Culture's Essential Role in Public Planning* (2001). Expressing the underlying assumption of sustainability, Hawkes accepts the premise that continuously pursuing "more material prosperity" "is not achievable either on a global scale or in the long term" (2001, 21). He recommends emphasizing "vitality" rather than "progress, development, or excellence" (2001, 23), for the latter three goals lead us back to unsustainable, losing outcomes if prioritized above all else. Hawkes writes: "In its simplest form, the concept of sustainability embodies a desire that future generations inherit a world at least as bountiful as the one we inhabit" (2001, 11). Unaware of Hawkes's writings, Muriel Meserve's reflection on the fishery and island closely resemble this concern: "There definitely is the idea of 'this is something we've been entrusted with and we need to carry forward to another generation'" (November 7, 2003). When describing the island's "rules," Muriel connects the rule about not picking too many flowers or digging them up to transplant elsewhere with a concern for others in the future: "You . . . don't take things away from [the island]; you leave it there. . . . You make it just as enjoyable for [the next person]" (October 5, 2009). The basic idea of conserving resources of all sorts for future generations quickly comes into conflict with a central myth of the modern era and of American culture in particular: that progress and development should be continually pursued. Sustainability, then, is a result of and a cog within *postmodern thought*, which questions the economic and environmental certainties of the late nineteenth and early to mid-twentieth centuries.

When one considers the power of the progress myth as a moral imperative separating modernity from postmodernity, it appears that pursuing sustainable practices is not just a new activity but also a *paradigm shift*, Thomas Kuhn's term for a radical change in the theoretical model for interpreting what is seen and influencing what is done (Bird 2011). Paradigm shifts do not happen neatly, completely, or all at once, and the work of reexamining and reinterpreting subjects from a new theoretical perspective after the initial paradigm shift has been called a "mopping-up" project, such as when feminist scholars reexamine familiar subjects to get a clearer idea of the status and behavior of females (Nielsen 1990, 21). Kuhn's theory began with physical science, the paradigm describing scientific perceptions, but in the social sciences and humanities, it is not just the researchers who shift their paradigms. When researchers and informants share a culture, the people who *live* the cultural systems the researchers study might change their paradigms at the same time the researchers do, changing their behavior and worldview and thus changing *culture*.

Because studying cultural change has far more complexity than the image of "mopping up" the surface of a floor, we need a different metaphor. Still a water metaphor, the image of a river works because of its depth and ability to flow and change direction when encountering obstacles. Initial "shifts" are akin to Kuhn's paradigm shift, but because cultural change happens through many people and their actions at different times, we need to consider the concept of "drift" as well.[2] Drifts occur when events are affected by initial paradigm shifts but (1) proceed in part from previous patterns, (2) gather momentum from their own movement, and/or (3) are influenced by other factors, and other shifts and drifts. There can be additional minor cultural shifts, like stones on the river bottom redirecting water flow, and there may be countershifts related to the term "backlash," which can affect the course or speed of cultural movement, but these do not necessarily fully negate a paradigm shift. Meanwhile, unrelated events in the course of history might also drop stones into the stream, again affecting individual lives and collective cultural change.

For example, the paradigm shift from patriarchy to gender equity has encouraged me as a feminist to look at how males and females work together at the Lewis Fishery in terms of sustaining the traditional practice of fishing and in terms of narrative stewardship.[3] In doing so, I observe the fact that women might or might not become full members of the crew, and some people might *see* women as full members of the crew while others might see them as ancillary or not see them at all. Whether a woman becomes a crew member depends on individual biography and other factors, such as local economics and shad population decreases, that operate concurrently with the gender paradigm shift. Together, the practical factors, shifts, and drifts produce the course of history. For instance, when the local mills closed, men had to go farther afield to find work, making it harder to fish on weeknights and making weekends more precious for home obligations. Meanwhile, the shad population dropped significantly, such that fisheries closed and shad fishing at the Lewis Fishery became less lucrative. Issuing from the paradigm of patriarchy, the gender wage gap together with the expectations that men will play the larger provider roles form the fishery's wider cultural context. These cultural norms operate as forces creating a smaller pool of men willing or able to spend their time in the increasingly unprofitable enterprise of shad fishing. Also drifting along from the patriarchal paradigm, women are still more likely to work without wages—activities traditionally known as "community service," "wifely duties," and "motherhood." So, as the fishery crew becomes an all-volunteer force, the economics more closely resemble women's work, making it more likely that women will become crew members.

This confluence of the new (gender equity) *and* old (patriarchal economics) paradigms enacted at once intersects with a transition from a material progress model to a sustainability model. However, whereas the gender paradigm shift that has been at work in the mainstream for generations is still being played out in shifts and drifts, the paradigm shift from progress to sustainability is relatively new. In general, sustainability as a strategy is most readily adopted when it takes the form of income-generating environmental efforts (marketing greenness) or efficient productivity (cutting costs by cutting hours or using less paper). It is more likely to be resisted when applying the concept threatens the myths of continuous economic and technological progress, which are integral parts of the American capitalistic worldview (e.g., using double-sided recyclable hard copies instead of individual electronic devices at a meeting may be dismissed as being "backward" even if the choice minimizes distraction, maximizes collaboration, and saves time). However, despite pressure (including device shaming) to unquestioningly adopt newer electronic technologies wherever possible, some people combine strategies for the best effect.

Embracing the idea of economic *sustainability* over economic *progress* often facilitates benefits related to other aspects of the sustainability paradigm shift (e.g., the environment, wellness, and community)—and this phenomenon is apparent on Lewis Island. The Lewis Fishery has sustained the haul seine tradition and supported the community for over 125 years by negotiating its way among what Hawkes calls the "4 pillars" of sustainability: "economic viability," "social equity," "environmental responsibility," and cultural sustainability, which he also calls "cultural vitality" (2001, 25). As it continues to be elsewhere in the United States, economic viability was once the king pillar at the Lewis Fishery, because additional income motivated crew members at shad fisheries on the Delaware from colonial times to the early twentieth century, when the economic fortunes of such fisheries took a beating from the drop in the shad population, loss of local train routes, and improvement in refrigeration (as Captain Fred Lewis told me in 1996, if you can eat nonlocal fish, you don't eat shad). By 1943, the other shad fisheries on the nontidal Delaware had disappeared, leaving the Lewis Fishery with the last remaining haul seining license, and in 1947 the IRS recategorized the fishery as a "hobby." As for being an economic pillar, because shad fishing is a seasonal endeavor, it has long been one among several strategies for local people to make a living, so crew members might not feel the shad fishery's economic losses quite so profoundly as workers in other fishing industries, even as Lewis Fishery shares dwindled and paid labor gradually switched over to being 100 percent volunteer labor. Yet, while profit is no longer a realistic aim, the Lewis Fishery cannot completely ignore the pillar of economic *viability* to remain economically sustainable.

During the twentieth and into the twenty-first century, environmental and cultural preservation have become entwined on Lewis Island, and the pillar of environmental sustainability has become the Lewis Fishery's ticket to economic *viability*, although not *profitability*. After Bill Lewis Sr. attracted government attention to the problem of the dropping shad population, the fishery's purpose remained serving its customers, but attention shifted from personal profit ever more to protecting the shad population and the river's health through studying the natural resources while keeping the operation at a modest size. The Big Story of the environment reflects a transformation of the economic model into an ecological one, starting with the story of Bill Lewis Sr. calling attention to the river's failing health, implicitly urging the dedication of funds to clean up the river.

Muriel Meserve describes this change from economic profit to environmental sustainability through her father Fred Lewis's contribution to educating the public about the needs of the fish:

> It was all about the fish ... which was always hard for a lot of them to understand[,] how he could care that much about these fish. I mean, it wasn't about ... a food source as much as about protecting the fish. And that was kind of a hard concept for a lot of people to wrap around, until they finally started doing the tagging [as part of an ecological study] and everything, and then it seemed easier for them to see that, you know, you were more interested in ... the information that you could get from them and not so much about how many you were going to sell on the bank, which became a really not even secondary, third, fourth, fifth, sixth something [laughs]. (October 5, 2009)

Muriel's son Steve continues the public education process and will often describe his role with a popular culture reference: "I liken myself to the Lorax." In the Dr. Seuss book *The Lorax* (1971) and later animated movie of that name (Pratt 1972), and the twenty-first century remake (Renaud and Balda 2012), the character the Lorax resists profitable economic activity to save the trees, saying that he "speaks for the trees," and Steve makes the comparison when he says that he "speaks for the fish." Pam Baker connects fishery activity to the larger competition between economics and environmentalism when she reflects, "And the river, you know, just like everything else, it's all important to keep everything healthy and in balance. And it's not okay to let these things die off and not know why. . . . [assuming an authoritative masculine voice] 'Maybe we should just let,' you know, 'it's too expensive.' . . . That doesn't seem quite right that we have a price tag on that" (October 3, 2009). While economics remains a concern, the goal is now to keep the fishery environmentally and financially sustainable, not necessarily profitable.

Over the years, the shad population has slowly but not steadily increased, with periods of dangerously low numbers punctuated by hope and ecstasy, such as in the 2013 season, when the catch was fourteen times larger than the season before, and again in the splashy 2017 season. In Hawkes's schema, this demonstrates success in the environmental pillar of sustainability, the pillar usually thought of first when anyone speaks of "sustainability" today. The two pillars—economic and environmental—were married at the Lewis Fishery in the 1970s and 1980s with a ten-year contract to tag fish, and then again in 2008, when Captain Steve worked out the contract with an intrastate cooperative to collect data and scale samples on an ongoing basis. While the earlier tagging operation earned good money for the crew (Ted Kroemmelbein, June 15, 2013), crew members receive no wages or shares under the current data collection contract. With *all* labor being volunteer labor, the modest fish sales plus the contract proceeds cover most of the fishery's regular expenses. The tagging contract also makes the fishery *legally* secure, for under it, the fishery has been released from its earlier requirement to catch and keep a certain number of fish in order to preserve the haul seine license. That security, in turn, makes it easier for the captain to continue two of the fishery's long-held environmental sustainability practices: (1) keeping just what's needed, and (2) throwing back small shad so that they will return as multiyear spawners.

From a public administration and economic and environmental sustainability standpoint, this collaboration is a huge success. The faithful tradition nicely fulfills the scientists' desire for consistent data collection, while the Lewis crew fishes the same way in the same place up to five nights each week during the season, as it has for over a century. Where else could the government find such a data-gathering operation so relatively inexpensively? Part of the data set already goes back to 1890, and for about four decades before the intrastate cooperative's contract, the fishery had voluntarily collected and provided other data and/or scale samples to the New Jersey Department of Environmental Protection. The one economic (efficiency) flaw in the solution is that the customary Saturday morning haul does not fall under the contract, which is for thirty evening hauls each season. This means that if the crew does not make a Saturday morning haul, there is little impact on the economic viability pillar. And in capitalistic American culture, where an accounting metaphor is used for gauging one's overall loss or gain, "the bottom line" is, well, literally the bottom line. Therefore, there is occasional grumbling from a couple of crew members about coming out for a Saturday morning haul, which tends to yield a low catch (and thus low/no sales) and which competes with Saturday morning chores and various community activities.

This Saturday morning pressure point in the new economic arrangement illustrates the importance of the cultural vitality pillar. At the Lewis Fishery, environmental and cultural preservation run in tandem because of some ubiquitous and all-encompassing aspects of the natural and cultural environments. That is, traditional haul seine fishing remains sustainable in Lambertville, New Jersey, in part because it straddles visibility and invisibility in relation to cultural sustainability. The Saturday morning haul is essential *not* for catching fish or fulfilling the terms of the new contract. Instead, it is essential because visitors, including potential customers, are more likely to be free on Saturday mornings to watch the haul, talk with others on the bank, and learn about the fishery, the island, and what they mean. Thus, the persistence of Saturday morning fishing is tightly intertwined with narrative stewardship—the system of telling symbolic Big Stories, microlegends, and everyday chitchat stories that together sustain sense of place and communal culture.

One could argue that the environmental and economic sustainability alignments at the fishery are tighter than the alignment with the social equity pillar in Hawkes's schema. If one considers race and ethnicity alone, one would notice that the crew and family are white, as one would expect given that the two surrounding counties, Hunterdon County, New Jersey, and Bucks County, Pennsylvania, are 91.1 percent and 89 percent white, respectively (US Census Bureau 2017), and the towns of Lambertville and New Hope are only slightly less so (CensusViewer 2017). While the crew's increasing diversity is in terms of gender, age, and full-time occupation, the increase in racial, ethnic, and religious diversity is in relation to customers and other visitors—arguably a trajectory that began in the first half of the twentieth century with the Lewis Fishery's efforts to fulfill the needs of Jewish customers who keep kosher. Since the 1960s, New Jersey has become home to immigrants from a far larger portion of the globe. Latino/Hispanic individuals now form just under 10 percent of the population of Lambertville, a statistic not broken down by race (CensusViewer 2017). This population is underrepresented among visitors, although some Latino children out with their Anglo friends and some Latino families do stop by the fishery in the evening. The underrepresentation is not surprising: shad is not a traditional food source in the Latino community, and during the fishing hours, members of this ethnic group are usually either working or visiting with each other at the park next to the school on the other side of town, which has come to function as a traditional "plaza" in Lambertville. Fishery crew and family interact with the Latino community at other times instead in their shared neighborhoods, the Methodist church, the library, and the schools; in businesses; and through friendship networks.

The Asian population, on the other hand, is statistically overrepresented, accounting for most of the customers on most nights but making up only 4 percent of the population of Hunterdon County, according to the 2017 US Census. This overrepresentation is not surprising given this group's reliance on shad for traditional culinary arts and the proximity to large Central Jersey Asian populations, just a thirty- to ninety-minute drive from Lambertville. In terms of race, religion, and ethnicity, social equity and sustainability intertwine in the ways that people from different cultural backgrounds connect with Lewis Island as befits their cultural traditions and individual biographies (e.g., whether they eat fish; the distance they live from Lambertville). Sense of place slowly evolves as the island "community" slowly changes with elective belonging. Distinct cultural traditions are sustained, and at the same time, Lambertville's culture of acceptance of difference is sustained.

Hawkes speaks at length to his audience of public planners about his preference for the term "cultural vitality" over "cultural sustainability," and the Lewis Fishery's relationships to culinary arts, narrative arts, and material culture exemplify why he carefully makes the distinction. Folklorists and cultural anthropologists define "culture" broadly to mean a shared, learned, and changing system of values, beliefs, behaviors, and objects created and used by a group of people—a definition that might embrace the social equity pillar, too. In public planning, however, "culture" is often defined as arts, and arts are primarily (but not exclusively) conceived according to the mainstream Western concepts that separate art (for its own sake) and craft (with some other use), that posit artists as specially trained practitioners, and that consider art objects and performances to be commodifiable, especially when they are "excellent" (Nanda and Warms 2010). "Excellence" does not seem to be all that important to the island's characteristic fishing activity, because haul seining requires a combination of simple tactile skills observers can easily grasp *and* advanced traditional river knowledge that is largely invisible. Also unseen is the traditional work of creating and maintaining island structures and fishery tools that give the place its traditional aesthetic. When the Lewis Fishery's culinary arts caught wide attention, it was through an appearance on an episode of the Travel Channel's *Bizarre Foods with Andrew Zimmern* (Zimmern 2015) and at the Smithsonian's Folklife Festival, cultural venues that do not need to narrowly define "art," or define "culture" as only art.

Because the public planners Hawkes addresses need to justify public funds allocation, the Western concepts of "commodities," "excellence," and "artists" as the most legitimate practitioners underpin criteria for which aspects of culture deserve the sustenance of public funding. However, these important concepts are largely beside the point in Lewis Island's communal life. The extended Lewis family is very involved in the performing arts professionally and personally, but this activity is usually done elsewhere, rather than on Lewis Island. On the island, people have been watching the hauls performed for free for generations as part of the sharing mandate most visitors do not know about. Lewis Island culture is most visible to the general public in connection with the Shad Fest, and, as pointed out earlier, many festival visitors assume that the fishery only operates when demonstrating shad hauling for the festival. Yet, these scheduled performances require no rehearsal or performance standards the crew must meet, as would be the case for mainstream artistic performances. Visual artists are welcomed on the island as part of the family's "mandate to share the island," but their presence is not essential to Lewis Island culture and usually happens at times other than the Shad Fest. Many consider the festival to be primarily a visual arts festival, although, like many community festivals (Rodríguez 1998; Lattanzi Shutika 2011), it has many contingent and contesting meanings. The Shad Fest's name and its tagline—a celebration of "the return of the shad"—sustain the festival's non-arts interpretations and alert festivalgoers and media audiences to the existence of Lewis Island, the Lewis Fishery, and local culture in the river community.

In short, Lewis Island culture exemplifies why Hawkes pointedly refuses to define "culture" as primarily art and emphasizes "vitality" rather than "excellence." Moreover, sharing the island at the festival is not only *part* of the island culture but also *just part of* the ongoing narrative stewardship system that makes the environmental, economic, social equity, and cultural vitality pillars functional.

At this point, I set aside the worthy question of how public planners might sustain culture to focus on how any of us can contribute to cultural sustainability, using narrative stewardship on Lewis Island as a model. The invisible embeddedness of Lewis Island's traditional culture has greatly contributed to sustaining communal traditions for over 125 years. With no motors on the boat, voices caught by the wind or muffled by the steady hum of cars over the steel grid bridge, dust-colored nets and footbridge, and just two buildings peeking through the foliage, Lewis Island does not attract attention. It is one of those places that is hidden in plain sight. Even if one has seen the Lewis Fishery at work, there's much more than meets the eye. It took me well over a decade to really *see* the fishery's system of narrative stewardship that combines Big Stories, character anecdotes, microlegends, and everyday storying to preserve the community resources of the shad, the river, traditional haul seine fishing, *and* the cultural practices of connecting with one another face-to-face through civility and caring. The story of fishery efforts to understand and preserve the shad population may make it into the regional newspapers, but crew member Sue Meserve's saga about building a theater set at work will stay largely invisible and embedded in fishery workings. Her story will not suffice on a storytelling stage, even though it interested other crew members for two weeks. Moreover, even if Sue's story were recognized through a publication, a grant, or contest winnings, it would likely not fulfill the cultural purpose of linking Sue with the others in her community. Nor would her entertaining a formal, conventional audience help Sue pass the time for herself and others on the crew, thus making their volunteer service rewarding, thus making fishing economically viable, and thus allowing the crew to provide the kind of fish needed in the Asian communities to support their foodways traditions in their homes. This trajectory from everyday storying to Indian foodways is one way the fishery interlinks with the social equity pillar of sustainability as well as the cultural vitality pillar in the form of traditional culinary arts. Again, much of this is invisible to most people.

Perhaps we can support cultural sustainability if we take from this case study some ideas about how visibility and invisibility can both empower and disempower those who are unseen. From social inequality theory, we learn that less powerful cultural groups are also less visible on the margins—as we have seen when cultural traditions move to the margins as changes in time, power, or migration put some other practice or people in the center. Moreover, many invisible traditions falter because they lack the visibility required to remind people to keep them going. However, inequality scholarship tells us that there is also power in invisibility, most obviously in the cultural coding described in Joan Radner and Susan Lanser's introductory essay to *Feminist Messages: Coding in Women's Folk Culture* (1993) and by James C. Scott in *Domination and the Arts of Resistance* (1990). Feminist scholarship suggests that there is power in being the *unmarked* category: the dominant group holds power because it is not marked as being unusual, but is the standard by which everything else is seen. For instance, maleness and whiteness are assumed until female or color identifiers appear. The dominant group's ethnicity or gender identity or other diversity category becomes invisible, because the dominant group's viewpoint is not looked *at* but rather *through*. This group's culture is the primary lens *through* which society at large looks, and actions that follow from the perception are considered to be natural. This is how hegemony works, after all: social inequalities reproduce themselves because the system fulfills the needs of the privileged using the contributions of the disadvantaged without anyone's overt, or even conscious, support. The system then reproduces itself with invisible seams, or even seamlessly.

Thus, sustainable efforts might be invisible or visible and still be effective. Some aspects of the fishery's sustainability have a debt to visibility's power—such as (1) the media coverage, and (2) a contract that links the Lewis Fishery to both government and science, powerful fields in America today. On the other hand, Lewis Island's traditions of civility and community are largely invisible because they are not joined by seams but are rather *interwoven* with and *embedded* in other parts of community culture. Even the usually visible economic viability pillar is partially embedded in the cultural valuing of volunteer service, which, like housewives' labor, is less visible than paid labor. Moreover, relationships across age diversity employ traditional nurturing-of-the-young and deference-to-age patterns to sustain the project in terms of both tradition and economics. Although Lewis Island is not an old-fashioned "children should be seen and not heard" locus, young and middle-aged adults solicit participation from and listen carefully to older adults, such as retired crew members, much like kids defer to their elders. Keziah notes that she defers to the adults more at the fishery than at her church community, and points out that she talks to the fishery captain and his wife more at family and friend gatherings than at the fishery when other adult male crew members besides her father are present. She provides this keen insight on the role of the captain: "I feel like it's more the adults', like, buddy-buddy time or whatever, 'cause it's like they're all friends together, and Steve kind of has to keep them happy and stuff because they're all volunteers . . . basically, so he kind of has to do that" (October 3, 2009). Here, she and the other kids pull back somewhat from being actors in the conversation, to help facilitate the economics of a volunteer operation. By moving to an audience role, or working or amusing themselves in some other way, the kids support the captain as he serves the social needs of the adult male crew members to keep the whole enterprise sustainable.

The enactment of narrative stewardship and civility extends throughout the fishing season, then into the rest of the year through other activities embedded in community life such as dog walking, sitting by the river, and attending religious services, which often go unnoticed in the wider community, although the island is in plain sight and accessible. Of course, the invisibility itself does *not* create Lewis Island's power, but the embeddedness of the Lewis Fishery and Lewis Island in local culture creates *both* the power of seamlessness *and* a sense of invisibility. In other words, the Lewis Fishery's sustainability depends in part on invisible efforts, but not on the invisibility of those efforts. Cultural sustainability persists in part because invisible individual commitments, not public planning, bridge the gaps between sustainable and unsustainable. So, the question now is how the lens of cultural sustainability and the discovery of the power of embeddedness should inform scholarship and community building.

I'm reminded of Jane Tompkins's essay "'Indians': Textualism, Morality, and the Problem of History" (1986), in which she considers how one could approach history after the advent of postmodern deconstruction. She finds that we still need to go about researching and writing history in much the same way, but with a heightened awareness of both new viewpoints and the increased tenuousness of any finding. Something similar can be said for scholars who study tradition. For a long time, scholars have looked askance at writings that simply celebrate traditions without applying critical thinking to analysis or interpretation. And yet, the existence of celebrational writings has had a hand in people in capitalistic societies committing individual resources to unprofitable cultural endeavors for the sake of tradition. At the same time, while embedded traditions are often more economically sustainable than those discrete traditions whose funding is dependent on public resources, cultural analysts who value social equity probably do not want to create unpaid opportunities for traditional cultural labor while public monies go elsewhere. Does that mean we should stop writing exhibit texts and full-length ethnographies? No. Just like the historians, we need to proceed with a new understanding of how academic and public sector study raises awareness of specific traditions and cultural sustainability itself. Detailed ethnographies show exactly how and where traditional culture is embedded but precarious, show why we need to expand public planning's concepts of culture beyond just arts and multiculturalism to include worldview and everyday life, to recognize and sustain, say, face-to-face "how was your day" conversations on a riverbank.

One can also pursue the personal benefits of cultural sustainability, including a sustained sense of place, in one's own lifetime, with one's own resources. Casey urges people to develop a "renewed sensitivity to place, affording a refreshed sense of its importance in our lives and those of others" to "offer a viable alternative to being and feeling out-of-place" (1993, 310).[4] Discussing the place-centered process of "homecoming," Casey observes that "homecoming, however much we may desire it, is not simply sweet but is often achieved with the most strenuous effort across the most daunting obstacles" (1993, 301). So too—and even more so—is the process of "re-implacement" that Casey propones. Cultural norms that associate capitalism and technology with myths of progress can pose such obstacles. Meanwhile, sense of place and narrative stewardship can provide resources for cultural sustainability. This takes awareness, agency, and commitment—including that "stubbornness" the Lewis family credits with preserving the haul seine tradition, the river, and the shad population. In a capitalistic culture, it is likely more difficult to choose cultural vitality over economic development, particularly since utilitarian ethics has assumed the authority of "common sense" in American mainstream culture, finding its application in aphorisms like "the bottom line is . . ." Ethical approaches other than utilitarianism—such as virtue ethics or ethics of kindness, both of which underpin sustainability and Lewis Island's Big Stories—are sometimes framed and dismissed as "idealistic" or "naïve."

If the profit paradigm is exchanged for a sustainability paradigm, however, other choices regarding time, place, and community emerge as appropriate, and the previous perpetual-progress cultural system is revealed as unrealistic or "naïve" in the context of limited local and global resources.

An important first step is toppling the capitalistic imperative of continual material gain and embracing instead economic viability. A second step is recognizing as myth (symbolic story) some of the cultural beliefs regarding electronic technology, such as the assumption that technologies inherently make tasks easier and more effective, together with the assumption that electronic technologies are the only technologies that "count" in comparison to other technologies such as pens, oars, needles, or wood stoves. The business world teems with examples of ineffective telectronics use, such as hearing someone snore on a conference call who would not have fallen asleep in a face-to-face meeting, unraveling conflicts arising from misunderstood abbreviations or tones in text messages, or spending hours loading updates or recovering from glitches caused by software programs marketed and released before they are ready. Moreover, many have found that the time properly working electronics can save is now spent working longer hours or addressing minutiae, because internet-enabled cell phones are assumed to make one available anytime and anyplace. In addition, expectations for performance have increased because people are expected to correct their errors with the available editing tools (which might themselves add errors with their autocorrect feature).

While global connections and long-distance family connections have become cliché in telectronics advertising, technology users subsequently become further *disconnected* from the friend or colleague actually in their physical presence (Turkle 2011, 2015), and from the places they are in—compounded by the stress of competing demands delivered electronically wherever and whenever without clear rules of etiquette or guidance for taking "time off" from electronic communication (Turkle 2011). Culturally and perhaps psychologically, continuous telectronic progress without humanistic intervention may not be a sustainable situation. Psychologists Miretta Prezza, Matilde Amici, Tiziana Roberti, and Gloria Tedeschi's findings may be applied here. Using American survey tools translated for an Italian audience, they assigned codes (for example, "Sense of Community") to the raw data and did statistical analysis to find that "Sense of Community is linked to Life Satisfaction and [inversely] to Loneliness even in a big city suburb" (Prezza et al. 2001, 48). Sometimes, people assume that a small, idyllic town is a requirement for sense of community to work, but Prezza and colleagues' study supports Casey's claim that sense of place has the power to combat anomie if people harness the power of telectronics rather than *being harnessed by* the perpetual dings and flashes demanding almost Pavlovian responses to attend to people in some other place (including across the room but separated by invisible microborders).

Of course, the answer is not in avoiding all electronics and other technologies but rather in claiming agency over them by committing to (1) critically thinking about progress and technology, (2) applying sustainability principles as appropriate, and (3) using sense of place and narrative stewardship as *cultural technologies*. Just as we should not destroy or abandon all technology unquestioningly, we shouldn't adopt it unquestioningly, either, and just such an intentional weighing process has evolved on Lewis Island. While personal benefit is still important, residents need to ask questions about the four pillars of sustainability, about quality of life: Is using a particular technology indeed actually easier, more effective, and more pleasant overall? Are place, time, and social needs met adequately? Just as with re-implacement, whereby people see places anew, we do not need to return to an earlier period of technology but rather to make a corrective shift—critically thinking about technology—to help orchestrate a drift across the complexity of social behavior and implaced experience to avoid the anomie that springs from rapid technological changes, social changes, or atopia (all three of which may accompany the unexamined use of communication technologies). One can build quality of life, sense of place, identity, and community that will sustain one by integrating a variety of electronic strategies *with* traditional strategies such as interacting in person, mending a net, or sharing a story, however mundane.

At the Lewis Fishery, technologies have been introduced slowly out of an acute awareness of the risk periodic flooding poses to expensive equipment, as well as out of respect for the cultural value of "just leave it . . . the way it is. . . . You know, it's good" (Muriel Meserve, October 5, 2009). It was even a noted change when Sue Meserve started using scissors rather than a knife to snip out the roe sacks without nicking them. On the other hand, methods used to build the footbridge and other structures have stayed more or less the same. Ted Kroemmelbein says with admiration: "Yeah, that whole building structure there, it's a testament to ingenuity. . . . It really is. You know, where they're up on piers, and the lower walls are designed to bust open and let the floodwaters through" (June 15, 2013). In contrast, outsiders often do not understand the importance of sustaining the island environmentally and sustaining the fishery economically by working *with* the natural forces of the river. When I describe a rebuilding operation, acquaintances will urge a technologically intensive solution that conquers nature and saves time. In the long run, most of the proposed ideas would likely cause more damage and lose more money—either through purchasing more-expensive equipment and materials, or by creating a jam on the creek that would push debris into other island structures and neighboring homes on the mainland, destroying them.

The old bridge-building method, however, preserves the wood *and* the activity of bridge building. Bridge building on the one hand, and social interaction, community building, and the narrative system on the other hand, mutually support one another. Power tools have certainly entered the equation, but the bridge structure, its functioning, and its construction process remain basically unchanged. These choices are in part aesthetic choices, as is evident when Sue Meserve describes rebuilding tasks following the big floods:

> It's always been done the same, and you get over there, and anything that you're gonna do, you do with reverence, because it always is the same. You know, it was a big deal to get the cabin rewired, and we kept impressing upon the guys who were doing it for us, even though they were friends of mine, it's like, "and don't go getting too fancy with it. Don't go, you know, messing it up so that it isn't the cabin it was." The same thing was thrown at the guy who redid the house recently. We didn't want to go too far with it. Even now, we think the rope light's really neat [that they themselves ran along the roof's edge], but we're going to take the rope light and put it up under the eaves so nobody can see it, so we get this nice glow, but we're not ruining the feeling of this . . . house built in the 1930s. (October 3, 2009)

Economic and environmental sustainability in the context of flooding also supports sustaining a particular set of aesthetic—cultural—values. Again, this is not simple nostalgia trying to restore the house back to the 1930s, but a continuance of the existing aesthetics, which to some degree has become a cherished part of the place.

Likewise on Lewis Island, people have adopted technologies for their practical usage and evaluated them according to the principles embedded in the Big Stories, and these adoptions may not match choices made in other venues. For example, the demands of their jobs compelled Sue Meserve (a manager and union officer), Steve Meserve (a senior computer programmer in a large corporation), and Dan Tuft (president of a disaster response company) to adopt cell phones early on. However, the way they use these tools on the island is closely monitored. Dan recalls talking about this with Steve back in the late 1990s:

> I remember the first time I used my cell phone there, before I was actually on the crew; I think, I was still [just] watching. And Steve made a comment about how he didn't think I was a person that used a cell phone that much. And I told him,

> I'm using the cell phone, actually, [because] it helps me come down to the island more, because I can still be in touch [with customers], but I can still be at the island too—because I was doing a work call. (October 3, 2009)

Initially, Sue and Steve only used their phones in the cabin before and after fishing to manage work duties, then left them in the drawer during a haul, as Dan still does. However, the growing Asian customer base understandably wanted to contact the fishery close to haul time to confirm whether there would be a haul before starting the long drive to the fishery after their commute from work. Sue and Steve then began to carry their phones, using the vibration setting to limit electronic disruptions in the valued natural setting.

Others on the crew generally leave phones up in the cabin during hauls, carving out time in this space for implacement and face-to-face community. While one youth new to fishing has been seen texting while putting on net, the longer-term youth leave their phones in the cabin or ignore them. Crew members Sarah Baker and Keziah Groth-Tuft clearly connect their understanding of careful technology use to the larger values of the place. When I asked Keziah to comment on the amount of cell phone usage on the island, she responded,

> It doesn't feel intrusive, 'cause at the island there's this sense that you know it's going to be changing, because people have other work—like that's why they don't fish all the time, because before they'd take off from the mill and stuff? But they don't do that now, so ... when you gotta be on your phone ... like people aren't texting all the time and ... whenever teenagers come down, it feels wrong, because like the whole point of it is like you learn how to talk to everyone *there*. (October 3, 2009)

Keziah's response shows an understanding that sustaining the tradition requires change, but that does not mean anything goes: those who are a consistent part of the operation set a tone such that a level of cell phone usage that would be normal elsewhere "feels wrong" on the island. Keziah's emphasis of place-oriented language is also noteworthy.

Like Keziah, Sarah Baker also appreciates having a place where communication technology use is minimized:

> The best thing is we just have phones, we don't have computers, we don't have TVs, you know, we don't have the internet, which is *crazy* now. And, in my generation—I'm nineteen—a lot of times teenagers ... you know, we're on Facebook, we're on Twitter, we're on Tumblr, but we don't *see* each other, you know,

face-to-face, and I think we've *lost* that through the years. I mean, don't get me wrong! I'm on Facebook, but I'm not *consumed* with it, you know. I *want* to see my friends; you know, I want to actually *go* out. And with fishing, you do, even if it is just three months. It's that three months of *connecting* with each other on like, on a daily basis, you know: how was your day? You know, actually *tell me* what you're *feeling*. (September 3, 2012)

Sarah and Keziah make the connection that technology is managed on the island to sustain other cultural norms of social interaction, which adds to the sense of the place. For Sarah, this seems to have an effect on the rest of her life, where she may participate with her peers, but she does not lose agency or become nullified, as suggested by her use of the word "consumed." Both Sarah and Keziah focus in on face-to-face conversation, Sarah specifically focusing on everyday storying as an activity worthy of the effort to sustain it, not just for themselves but for others in their generation.

It is indeed too early for researchers and policy makers to give up face-to-face communities as a thing of the past, but at the same time, these become increasingly special, and so do face-to-face narrative situations. Anthropologist Julie Cruikshank writes,

> In northern Canada, storytellers of Yukon First Nations ancestry continue to tell stories that make meaningful connections and provide order and continuity in a rapidly changing world. An enduring value of informal storytelling is its power to subvert official orthodoxies and to challenge conventional ways of thinking. (1998, xiii)

The everyday storying Sarah refers to seems simple, but its power is that almost *anyone* can do it, for it does not require specific artistic skill, electronics, or an account with a corporation. Becoming less frequent but by no means rare, face-to-face narrative stewardship subverts the growing conventions of teenage social media usage as well as the power of the telectronics industry. This orally originating content is effective even when those stories are disseminated through the electronic media, as with blogs, radio, television, periodicals, and personal social media accounts of visitors, crew members, and the fishery family. We don't have to be Luddites *or* techno-lemmings; there is plenty of room in the middle ground.

Narrative stewardship is indeed a strategy for cultural sustainability, for guarding against anomie and postmodern alienation. Similar to the situation in the Yukon Territory, the content of narrative stewardship on Lewis Island "make[s] meaningful connections and provide[s] order and continuity in a rapidly changing world." Edward Casey reassures us that "the existence of pictorial and narrational journeys to and between places reminds us that we are not altogether without resources in our placelessness" (1993, 310), and narrative is indeed one of those resources. Surely not the only tool or a fix-all in a monolithic trajectory, narrative stewardship is a way to "put an oar in" to navigate oneself and others across the shifts and drifts of cultural change—even reclaiming agency in midstream and influencing the course or nature of cultural change, to a greater or lesser extent. Yet, while one may crave agency as power, wielding individual power is not easy given the complexity of organic culture and the complexity of narrative stewardship under examination in this ethnography.

At the Lewis Fishery, the narrative system re-implaces people through both the stories and the activity of telling stories at that site in that social situation. The process encourages both implacement and community building, especially as they relate to the Big Story of civility. Friendly behaviors are encouraged in this place, reflected and enacted through narration, giving to participants the agency to build sense of place and combat feelings of alienation. Robert Archibald argues that "the past and its stories are a priority. They remind us of the continuity of time and provide an antidote to the isolation of the present" (2004, 35). In Archibald's view, stories bring together the concepts of time and space (which create isolation when separated) for the sake of emotional wellness. I would add that the process of telling these stories has a major impact on both tellers and listeners, who together take part in narrative stewardship. Barbara Johnstone reminds us that with American personal narratives and with "myths, tales, and stories told in all Western cultures ... storytellers must create a sense of rising tension as they narrate, and this tension must be released at the end of the narrative core" (1990, 34). On Lewis Island, this common American narrative pattern encourages the success of the Big Stories through the combination of narrative content and process. That is, the environmental Big Story of the uncertain shad population told on the bank raises that tension, and the people watching the fishing and scientific data collection can see that something is being done about the concern. Thus, the ritual visit resolves some of that tension, making effective narration, while also leaving enough tension to motivate listeners to attend to environmental issues themselves.

To create sense of place and sense of belonging, it appears that narrative stewardship benefits from tellers and audiences being in the same place, even as it also uses some telectronic technologies to good effect. Being together takes commitment, like the commitment and agency of elective belonging, like the commitment to the fishery's routine work and the island's routine maintenance. Steve Meserve aptly sums up the challenges:

> Running the fishery, it's certainly as tough as I ever thought it would be. I mean, keeping a crew together is not an easy thing. The work is hard, the pay is minimal even in good years, and not catching fish is disheartening. I mean, when we have a good haul of fish, everyone feels a whole lot better. You get a few nights with no fish at all, if the water's high or something. That's probably the hardest time to be fishing. It's cold, it's wet, and there are no fish. What are we doing out here? So, to try to keep people engaged and all, and add to it the stresses of everyday life, you know, going to work every day and working those hours and *then* doing this, and not being able to—as my grandfather was at the end of his participation—you know, he would spend all day mending so that we could just go and fish. That's a luxury we don't have right now. (October 3, 2009)

While the enterprise becomes more difficult, Steve has a heritage of "stubborn" captains, crews, and family members to draw on to sustain the activity and the place. His family and crew also have the resource of narrative stewardship, which not only sustains the island, river, fish, tradition, and community but also provides a way to keep people engaged and help them relieve "the stresses of everyday life" by going to "a whole 'nother place."

Narrative stewardship requires some degree of commitment on the part of individuals and communities, but it is also a valuable resource that has the potential for wide application, embedded in everyday life. Through their studies, folklorists, anthropologists, sociologists, cultural geographers, and historians can make visible the power of using narrative to steward community resources as part of cultural sustainability. Such studies reveal the drifts in cultural currents that accompany the sometimes confluent and sometimes conflicting paradigm shifts of technology and sustainability. As individuals, we can steward our own valued resources with more awareness. The solution is as simple as going someplace and engaging in storytelling, but then again ... it's not. Just a few stories won't do the trick—although that's the start of the process. Rather, sustenance comes from commitment to pervasive engagement and repetition—on the part of the group and of the individuals who make up the community. On Lewis Island, commitment to cultural sustainability is visible on yet another night when supper is late for folks who smell of shad and river. It happens in this place with another sunset, another story, another haul.

APPENDIX A

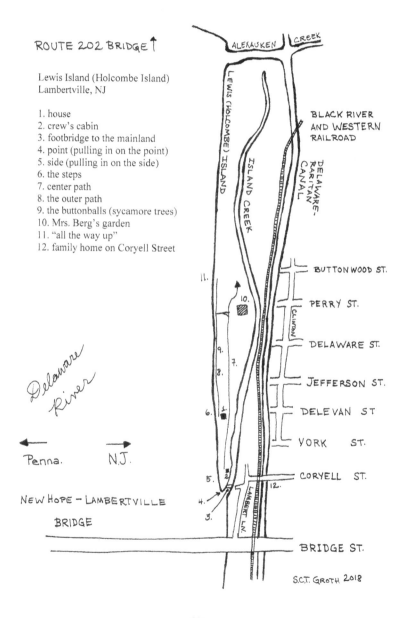

Lewis Island (Holcombe Island)
Lambertville, NJ

1. house
2. crew's cabin
3. footbridge to the mainland
4. point (pulling in on the point)
5. side (pulling in on the side)
6. the steps
7. center path
8. the outer path
9. the buttonballs (sycamore trees)
10. Mrs. Berg's garden
11. "all the way up"
12. family home on Coryell Street

APPENDIX B

A Partial Representation of the Extended Lewis Family

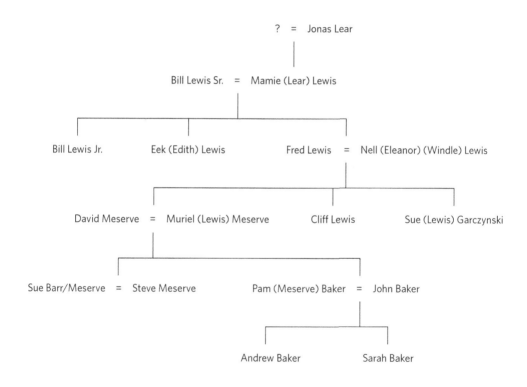

APPENDIX C

Selected Lewis Fishery Statistics (1890–2017)

Year	Total Catch*	Days Fished	Total Hauls
1890	3500		
1891	2500		
1892	1800		
1893	1200		
1894	1600		
1895	4028		
1895	9288		
1896	9288		
1897	2500		
1898	2000		
1899	1800		
1900	2000		
1901	2200		
1902	1500		
1903	1900		
1904	1600		
1905	1400		
1906	1300		
1907	1700		
1908	1400		
1909	3000		
1910	5923		
1911	2800		
1912	5749		
1913	4972		
1914	1600		

Year	Total Catch*	Days Fished	Total Hauls
1915	300		
1916	700		
1917	900		
1918	400		
1919	3500		
1920	450		
1921	1035		
1922	1466		
1923	1648		
1924	3755		
1925	742	42	458
1926	661	31	208
1927	1061	46	436
1928	2174	53	543
1929	2706	64	616
1930	470	45	362
1931	887	53	501
1932	1442	47	450
1933	2325	45	420
1934	1796	48	520
1935	4417	42	328
1936	951	45	420
1937	4161	32	448
1938	3240	63	693
1939	4439	48	506
1940	611	29	170
1941	129	30	162
1942	1096	39	193
1943	3025	44	215
1944	226	15	45
1945	295	37	144
1946	254	31	118
1947	1358	46	358
1948	43	28	59
1949	3	30	32
1950	9	41	51

Appendix C

Year	Total Catch*	Days Fished	Total Hauls
1951	25	34	38
1952	27	36	43
1953	0	30	31
1954	9	19	26
1955	36	40	43
1956	0	29	32
1957	10	9	12
1958	54	14	18
1959	27	16	24
1960	6	12	19
1961	90	14	26
1962	250	15	18
1963	3983	40	70
1964	1646	39	90
1965	319	31	48
1966	77	38	44
1967	243	35	65
1968	33	22	27
1969	90	18	29
1970	122	15	25
1971	664	30	54
1972	348	30	64
1973	496	32	69
1974	417	29	49
1975	1738	38	117
1976	1470	38	123
1977	1120	37	110
1978	1226	40	121
1979	2003	38	107
1980	1920	44	148
1981	6392	52	118
1982	3789	42	127
1983	1444	32	100
1984	2383	38	152
1985	2202	41	69
1986	3036	41	99

Year	Total Catch*	Days Fished	Total Hauls
1987	1830	44	111
1988	2778	49	78
1989	4646	41	89
1990	2332	49	92
1991	2312	49	76
1992	4790	44	94
1993	347	33	36
1994	387	26	49
1995	1257	43	66
1996	209	34	57
1997	550	39	46
1998	939	30	49
1999	198	34	43
2000	183	37	45
2001	219	26	32
2002	469	40	52
2003	474	38	56
2004	292	46	55
2005	104	33	36
2006	73	42	44
2007	66	20	21
2008	83	37	38
2009	112	43	44
2010	489	35	38
2011	52	24	25
2012	197	35	36
2013	1015	36	37
2014	386	33	36
2015	329	35	38
2016	369	37	41
2017	1262	34	43

*In earlier years, counts were often rounded to even figures.
Source: Stephen Meserve, Lewis Fishery

NOTES

Preface

1. This chapter borrows heavily from two of my American Folklore Society conference papers, "Staring Down the Ghosts of Going Native; or, Professor, Can I Say 'I'?" (2010) and "Fishing in the Mainstream: (In)visibility, Embeddedness, and Sustainability" (2013).

Chapter 1. Welcome to the Island: The Lewis Fishery in Context

1. The 1967 season alone boasted performances by the Lovin' Spoonful, Ian and Sylvia, Judy Collins, Dave Brubeck, Simon and Garfunkel, the Four Seasons, Paul Anka, Buddy Rich, the Jefferson Airplane, Lou Rawls, Victor Borge, Flatt and Scruggs, Robert Goulet, Woody Allen, the Four Tops, and Buffy Sainte-Marie (Case 2005–2016). For an illustrated history of Lambertville, see Mastrich, Warren, and Kline (1996) and Warren and Toboz (1998).

Chapter 2. Fishing with Purpose: The Big Stories

1. The word Pam's trying to recall here is "stubbornness," which Muriel Meserve usually uses, and which is discussed elsewhere in this book.

2. This is not to say that Lambertville is the only place to have a shad-oriented event; there are many others. C. Boyd Pfeiffer remarks that shad festivals were common along the Hudson River at the turn of nineteenth century and into the twentieth (1975, 156). The bipartisan albeit now Republican-heavy Shad Planking in Virginia began in the 1930s to mark the James River shad run, and it continues on as more of a political event (Vozzella 2016; Ottenhoff 2017). Both Essex, Connecticut's, shad festival and the Bethlehem Shad Festival in Bethlehem, Pennsylvania, started in 1978 (according to the Delaware River Shad Fishermen's Association website), before the Lambertville one finally came to fruition, and ran for a couple decades. The Fishtown neighborhood of Philadelphia started a shad festival in 2009.

3. Steve and Sue Meserve note that when the no-beer-in-the-fridge rule was finally broken on the island, drinking was still very limited, a keg's worth lasting for the entire season (April 22, 1996).

4. This sign has been reproduced several times (it's often taken by floodwaters), either with or without the word "the," and sometimes contracting "do not" to "don't."

5. For examples of articles from higher-education publications that trace history, examine challenges, and offer solutions to what is seen as a widespread problem with lack of civility on college campuses, see Connelly (2009), Gilroy (2008), and Levine (2010). Porath, MacInnis, and Folkes (2010) and Oore et al. (2010) offer scholarly research on civility in the workplace, while Hauser (2011) and Kear (2011) are examples of articles in industry publications that summarize research and inform practice.

6. The implication here is swear words.

7. The missionaries, Muriel Meserve explains, were working a street ministry in New Hope during the civil turbulence of the 1960s and 1970s, when youth who were attempting to "find themselves" instead lost themselves and needed help getting out of various troubles (personal communication, June 27, 2017).

Chapter 3. The Captains: Between Myth and Legend, Article and Anecdote

1. Different storytellers set this incident in either Philadelphia or Trenton, referring to both as "down there."

Chapter 4. "Were You There When . . . ?": Microlegends

1. For visuals of the "Great Flood," see *The Great Flood of 1955: Picture Story of the Greatest Catastrophe in the History of Hunterdon and Bucks Counties* (n.d.).

Chapter 5. "It's Like I Said to So-and-So": Everyday Storying

1. Portions of the discussion on dogs were presented in Groth, "Stories Unleashed: Dogs' Role in Narrative Stewardship on Lewis Island" (2015).

Chapter 6. Talking the Walk: Processional Storytelling

1. See Hufford 1992, 163ff., for her Turnerian analysis of structure and anti-structure.

2. See Turner (1969) 1995 for the classic treatment of structure and anti-structure.

3. See Turner (1969) 1995 for the theory of "liminality," the "betwixt and between" stage within rites of passage.

4. See Hufford 1992 and Geertz 1973a for similar connections between cultural texts and understanding oneself and one's position in existence.

Chapter 7. Who-All's Coming Down to the Island: Belonging at the Lewis Fishery

1. A much-expanded version of this section was earlier presented in a paper entitled "Chick Hauls and Fishwives: Continuity and Change in a Traditional Haul Seine Fishery" (Groth 2012).

Chapter 8. "A Whole 'Nother Place": Narrative Stewardship and Sense of Place

1. Muriel Meserve recalls that her mother actually ran the pop stand during the years when town residents would come to swim off the island, and she and Cliff simply helped. The bottles were collected and returned to the bottling company in New Hope. Drink sales resumed decades later when Muriel and her fellow Lambertville Public School teachers organized it as a fundraiser during the Shad Festival (personal communication, spring 2016). With multiple projects and floods moving objects and topsoil, it makes sense that an assortment of bottles clustered on the island, but readers should not get the impression of a garbage dump. While slightly faulty in the details, Sue Meserve is correct to observe that the island's history as a social place remains apparent in the landscape.

2. Although "emplacement" is the more common word form, I will follow Casey's use of "implacement" and "re-implacement"), not only because I am referring to his theory but also because "emplacement" is more commonly used to refer to a situation, and both Casey and I are often emphasizing not the place itself but the process of something being put into a place.

3. Here I must share an anonymous reviewer's astute observation that this relationship between place and self in contemplation may explain why resolution comes to troubled visitors when they simply walk on the island or watch the river from the bank.

Chapter 9. Fishing in the Mainstream: Anomie, Sustainability, and Narrative Stewardship

1. A significant portion of this chapter was first presented in a paper entitled "Fishing in the Mainstream: (In)visibility, Embeddedness, and Sustainability" (Groth 2013).

2. The concept of "shift and drift" developed in large part through conversations about feminism and cultural change between folklorist Margaret Mills and myself.

3. This example owes much to my 2012 American Folklore Society conference paper entitled "Chick Hauls and Fishwives: Continuity and Change in a Traditional Haul Seine Fishery" (Groth 2012).

4. Casey's use of "out-of-place" in this context should not be confused with the admonition to "know one's place." This second usage usually relates to knowing one's role in a situation and is often used to pressure people in lower positions in stratified systems (e.g., females, blacks, children) to stay within their lower-status roles. Here, Casey and I are

addressing human relationship with place more generally (as opposed to being generally *un*attached to places and spaces, in a state of anomie). To be sure, revealing how specific places are used to preserve stratified social roles also needs more attention from researchers as a scholarly "mopping-up" activity related to making the paradigm shift from social inequity to social equity.

REFERENCES

Albert, Richard C. 2002. "In-Tocks-icated: The Tocks Island Dam Project." *CRM: Cultural Resource Management* 25(3): 5–8. At http://www.nps.gov/history/crmjournal/CRM/v25n3.pdf.

Alleyne, Brian. 2002. "An Idea of Community and Its Discontents: Towards a More Reflexive Sense of Belonging in Multicultural Britain." *Ethnic and Racial Studies* 25(4): 607–27.

Alper, Becka A., and Daniel V. A. Olson. 2011. "Do Jews Feel Like Outsiders in America? The Impact of Anti-Semitism, Friendships, and Religious Geography." *Journal for the Scientific Study of Religion* 50(4): 822–30. doi:10.1111/j.1468-5906.2011.01599.x.

Alter, Adam. 2017. *Irresistible: The Rise of Addictive Technology and the Business of Keeping Us Hooked*. New York: Penguin Press.

Anderson, Benedict. 1991. *Imagined Communities: Reflections on the Origin and Spread of Nationalism*. Rev. ed. London: Verso.

Archibald, Robert R. 2004. *The New Town Square: Museums and Communities in Transition*. Walnut Creek, CA: AltaMira Press.

Avnet, Jon, dir. 1991. *Fried Green Tomatoes*. Universal Pictures.

Bascom, William. 1983. "Malinowski's Contributions to the Study of Folklore." *Folklore* 94(2): 163–72. At http://www.jstor.org/stable/1260489.

Basso, Keith. 1984. "'Stalking with Stories': Names, Places, and Moral Narratives among the Western Apache." In *Text, Play, and Story: The Construction and Reconstruction of Self and Society*, edited by Keith Basso, 19–55. Washington, DC: American Ethnological Society.

Bauman, Richard. 1972. "The La Have Island General Store: Sociability and Verbal Art in a Nova Scotia Community." *Journal of American Folklore* 85(338): 330–43.

Bauman, Sheri, Russell B. Toomey, and Jenny L. Walker. 2013. "Associations among Bullying, Cyberbullying, and Suicide in High School Students." *Journal of Adolescence* 36(2): 341–50. doi:10.1016/j.adolescence.2012.12.001.

Becker, Mark W., Reem Alzahabi, and Christopher J. Hopwood. 2013. "Media Multitasking Is Associated with Symptoms of Depression and Social Anxiety." *Cyberpsychology, Behavior, and Social Networking* 16(2): 132–35. doi:10.1089/cyber.2012.0291.

Belanus, Betty J. 2004. "Water Ways: Charting a Future for Mid-Atlantic Maritime Communities." In *Smithsonian Folklife Festival*, 56–71. Washington, DC: Smithsonian Institution.

Ben-Amos, Dan. 1992. "Folktale." In *Folklore, Cultural Performances, and Popular Entertainments*, edited by Richard Bauman, 101–18. New York: Oxford University Press.

Bendix, Regina. 1997. *In Search of Authenticity: The Formation of Folklore Studies*. Madison: University of Wisconsin Press.
Bird, Alexander. 2011. "Thomas Kuhn." In *The Stanford Encyclopedia of Philosophy*. Center for Study of Language and Information, Stanford University. At http://plato.stanford.edu/entries/thomas-kuhn/.
Booth, Alison, and Kelly J. Mays, eds. 2011. *The Norton Introduction to Literature*. Portable ed. New York: W. W. Norton.
Bowman, Marion. 2004. "Procession and Possession in Glastonbury: Change and the Manipulation of Tradition." *Folklore* 115(3): 273–85. At http://www.jstor.org/stable/30035212.
Bruhn, John G. 2005. *The Sociology of Community Connections*. New York: Springer.
Bueker, Catherine Simpson. 2013. "'Leads' to Expanded Social Networks, Increased Civic Engagement and Divisions within a Community: The Role of Dogs." *Journal of Sociology and Social Welfare* 40(4) (December): 211–36. Academic Search Premier (92611548).
Burnett, Frances Hodgson. (1911) 1962. *The Secret Garden*. Philadelphia: J. B. Lippincott.
Case, Jon, comp. 2005–2016. "The 1967 Season." St. John Terrell's Lambertville Music Circus Summer Stock Theater. At http://www.lambertville-music-circus.org/concerts/1967.html. Accessed July 10, 2015.
Casey, Edward S. 1993. *Getting Back into Place: Toward a Renewed Understanding of the Place-World*. Bloomington: Indiana University Press.
Cashman, Ray. 2008. *Storytelling on the Northern Irish Border*. Bloomington: Indiana University Press.
Cashman, Ray. 2011. "Text and Community Forum: *Storytelling on the Northern Irish Border*." Panel discussion at the annual meeting of the American Folklore Society, Bloomington, Indiana, October 13.
Castagna, Michael de Freitas. 1999–2002. "A Brief History of Finkles Hardware." At http://www.finkles.com/. Accessed June 24, 2014.
CensusViewer. 2017. "Lambertville, New Jersey Population: Census 2010 and 2000 Interactive Map, Demographics, Statistics, Quick Facts." At http://censusviewer.com/city/NJ/Lambertville. Accessed June 28, 2017.
Chesley, Noelle. 2005. "Blurring Boundaries? Linking Technology Use, Spillover, Individual Distress, and Family Satisfaction." *Journal of Marriage and Family* 67(5): 1237–48. doi:10.1111/j.1741-3737.2005.00213.x.
Cockshaw, Wendell D., Ian M. Shochet, and Patricia L. Obst. 2014. "Depression and Belongingness in General and Workplace Contexts: A Cross-Longitudinal Investigation." *Journal of Social and Clinical Psychology* 33(5): 448–62. doi:10.1521/jscp.2014.33.5.448.
Cohen, Elisia L., Sandra J. Ball-Rokeach, Joo-Young Jung, and Yong-Chan Kim. 2002. "Civic Actions after September 11: Exploring the Role of Multi-Level Storytelling." *Prometheus* 20(3): 221–28. doi:10.1080/08109020210141344.
Connelly, Robert J. 2009. "Introducing a Culture of Civility in First-Year College Classes." *Journal of General Education* 58(1): 47–64. ERIC (EJ850401).
Cruikshank, Julie. 1998. *The Social Life of Stories: Narrative and Knowledge in the Yukon Territory*. Lincoln: University of Nebraska Press.

D'Autrechy, Phyllis B. 1993. *Hunterdon County New Jersey Fisheries, 1819–1820*. Flemington, NJ: Hunterdon County Historical Society.

Dégh, Linda. 1995. *Narratives in Society: A Performer-Centered Study of Narration*. Helsinki: Academia Scientiarum Fennica.

Delaware and Raritan Canal Commission. 2012. "Delaware and Raritan Canal History." Delaware and Raritan Canal State Park. At http://www.dandrcanal.com/history.html. Accessed August 20, 2016.

Delaware River Shad Fishermen's Association. 2016. At http://www.drsfa.org/. Accessed January 5, 2016.

Dorson, Richard. 1973. "Mythology and Folklore." *Annual Review of Anthropology* 2: 107–26. At http://www.jstor.org/stable/2949263.

Dougherty, Kevin D., and Andrew L. Whitehead. 2011. "A Place to Belong: Small Group Involvement in Religious Congregations." *Sociology of Religion* 72(1): 91–111. doi:10.1093/socrel/srq067.

Dr. Seuss [Theodor Seuss Geisel]. 1971. *The Lorax*. New York: Random House.

Duffy, Erin. 2011. "N.Y. Reservoir Changes Could Ease Flooding Concerns for N.J.'s Delaware River Towns." *Times of Trenton*, April 8. At http://www.nj.com/mercer/index.ssf/2011/04/ny_reservoir_changes_could_eas.html.

Dunaj, Greg. 2009. "ROE!" May 6–June 9. At http://shad-roe.blogspot.com/. Accessed June 12, 2014.

Edwards, Viv, and Thomas J. Sienkewicz. 1990. *Oral Cultures Past and Present: Rappin' and Homer*. Oxford: Basil Blackwell.

Encyclopædia Britannica. 2016. S.v. "anomie." Last modified, November 26, 2014. At http://www.britannica.com/topic/anomie.

Evers, Larry, and Barre Toelken. 2001. "Collaboration in the Translation and Interpretation of Native American Oral Traditions: Introduction." In *Native American Oral Traditions: Collaboration and Interpretation*, edited by Larry Evers and Barre Toelken, 1–14. Logan: Utah State University Press.

Gallagher, Sarah A. 1903. *Early History of Lambertville, NJ*. Trenton, NJ: MacCrellish and Quigley.

Garber, Daniel. 2018. *Lambertville Beach*. James A. Michener Art Museum, Doylestown, Pennsylvania. At https://bucksco.michenerartmuseum.org/bucksartists/image/210/. Accessed June 17, 2018.

Geertz, Clifford. 1973a. "Deep Play: Notes on a Balinese Cock Fight." In *The Interpretation of Cultures: Selected Essays*, edited by Clifford Geertz, 412–53. New York: Basic Books.

Geertz, Clifford. 1973b. "Ethos, Worldview, and the Analysis of Sacred Symbols." In *The Interpretation of Cultures: Selected Essays*, edited by Clifford Geertz, 126–41. New York: Basic Books.

Geertz, Clifford. 1973c. "Thick Description: Towards an Interpretive Theory of Culture." In *The Interpretation of Cultures: Selected Essays*, edited by Clifford Geertz, 3–30. New York: Basic Books.

Gillen-O'Neel, Cari, and Andrew Fuligni. 2013. "A Longitudinal Study of School Belonging and Academic Motivation across High School." *Child Development* 84(2): 678–92. doi:10.1111/j.1467-8624.2012.01862.x.

Gilroy, Marilyn. 2008. "Colleges Grappling with Incivility." *Education Digest* 74(4): 36–40. ERIC (EJ888633).
Girl Scouts of the United States of America. 2018. "Spirituality/Religion." In *Blue Book of Basic Documents 2018*, 24. New York: Girl Scouts of the United States of America.
Giuffo, John. 2013. "America's Prettiest Towns." *Forbes*, August 16.
Glassie, Henry. 1982. *Passing the Time in Ballymenone*. Philadelphia: University of Pennsylvania Press.
Glassie, Henry. 1983. *All Silver and No Brass: An Irish Christmas Mumming*. Philadelphia: University of Pennsylvania Press.
Goffman, Erving. 1959. *The Presentation of Self in Everyday Life*. New York: Doubleday.
Goode, Daniel J., Edward H. Koerkle, Joan D. Klipsch, and Amy L. Shallcross. 2010. "Development of a Flood-Analysis Model for the Delaware River." Paper presented at the Second Joint Federal Interagency Conference, Las Vegas, Nevada, June 27–July 1.
The Great Flood of 1955: Picture Story of the Greatest Catastrophe in the History of Hunterdon and Bucks Counties. n.d. Flemington, NJ: Democrat Press.
Groth, Susan Charles. 1998. *Cultural Riches: A Sampler of Northwest Jersey Traditional Arts in Cultural Context*. Oxford, NJ: Warren County Cultural and Heritage Commission.
Groth, Charlie. 2010. "Staring Down the Ghosts of Going Native; or, Professor, Can I Say 'I'?" Paper presented at the annual meeting of the American Folklore Society, Nashville, Tennessee, October 14.
Groth, Charlie. 2011. "Covering Lewis Island: Media's Role in Narrative Stewardship at a Traditional Fishery." Paper presented at the annual meeting of the American Folklore Society, Bloomington, Indiana, October 13.
Groth, Charlie. 2012. "Chick Hauls and Fishwives: Continuity and Change in a Traditional Haul Seine Fishery." Paper presented at the annual meeting of the American Folklore Society, New Orleans, October 25.
Groth, Charlie. 2013. "Fishing in the Mainstream: (In)visibility, Embeddedness, and Sustainability." Paper presented at the annual meeting of the American Folklore Society, Providence, Rhode Island, October 17.
Groth, Charlie. 2015. "Stories Unleashed: Dogs' Role in Narrative Stewardship on Lewis Island." Paper presented at the annual meeting of the American Folklore Society, Long Beach, California, October 15.
Groth, Susan Charles. 1998. *Cultural Riches: A Sampler of Northwest Jersey Traditional Arts in Cultural Context*. Oxford, NJ: Warren County Cultural and Heritage Commission.
Gutierrez, Peter M., Lisa A. Brenner, Jeffrey A. Rings, Maria D. Devore, Patricia J. Kelly, Pamela J. Staves, Caroline M. Kelly, and Mark S. Kaplan. 2013. "A Qualitative Description of Female Veterans' Deployment-Related Experiences and Potential Suicide Risk Factors." *Journal of Clinical Psychology* 69(9): 923–35. doi:10.1002/jclp.21997.
Hall, Edward. 2010. "Spaces of Social Inclusion and Belonging for People with Intellectual Disabilities." *Journal of Intellectual Disability Research* 54 (supp. 1): 48–57. doi:10.1111/j.1365-2788.2009.01237.x.
Hardin, Garrett. (1968) 2005. "The Tragedy of the Commons." Garrett Hardin Society. At http://www.garretthardinsociety.org/articles/art_tragedy_of_the_commons.html.

Harrington, M. R. 1938. *The Indians of New Jersey: Dickon Among the Lenapes.* New Brunswick, NJ: Rutgers University Press.
Hauser, Susan. 2011. "The Degeneration of Decorum." *Workforce Management* 90(1): 16–21. Academic Search Premier (57678806).
Hawkes, Jon. 2001. *The Fourth Pillar of Sustainability: Culture's Essential Role in Public Planning.* Melbourne: Cultural Development Network.
Hazen, Joseph. 1998. Chit-Chat. *Lambertville Beacon,* April 2.
Hobsbawm, Eric, and Terence Ranger, eds. 1983. *The Invention of Tradition.* Cambridge: Cambridge University Press.
Hoffman, Adria Rachel. 2012. "Exclusion, Engagement, and Identity Construction in a Socioeconomically Diverse Middle School Wind Band Classroom." *Music Education Research* 14(2): 209–26. doi:10.1080/14613808.2012.685452.
Hufford, Mary T. 1992. *Chaseworld: Foxhunting and Storytelling in New Jersey's Pine Barrens.* Philadelphia: University of Pennsylvania Press.
Hymes, Dell. 1975. "Breakthrough into Performance." In *Folklore: Performance and Communication,* edited by Dan Ben-Amos and Kenneth S. Goldstein, 11–74. The Hague: Mouton.
Jackson, Patrick. 2012. "Situated Activities in a Dog Park: Identity and Conflict in a Human-Animal Space." Society and Animals 20: 254–72. At http://pacificrime.org/jackson2012as.pdf.
James A. Michener Art Museum. 2018. "Daniel Garber." At https://bucksco.michenerartmuseum.org/bucksartists/artist/83/. Accessed June 17, 2018.
Johnson, Lady Bird. 1970. *A White House Diary.* New York: Holt, Rinehart and Winston.
Johnstone, Barbara. 1990. *Stories, Community, and Place: Narratives from Middle America.* Bloomington: Indiana University Press.
Joselit, Jenna Weissman. 2013. "Playing Jewish Geography from California to the New York Islands: Exhibit Shows Differences between West and East Coast Jews." *Jewish Daily Forward,* April 10. At http://forward.com/articles/174188.
Katz, James E. 2006. "Mobile Communication and the Transformation of Daily Life: The Next Phase of Research on Mobiles." *Knowledge, Technology, and Policy* 19(1): 63–71. Academic Search Premier (23303042).
Kear, Mavra. 2011 "Speak Out for Civility." *Florida Nurse* 59(1): 16. Academic Search Premier (59528190).
Kehilat HaNahar. 2015. "About." At https://kehilathanahar.org/content/about. Accessed December 15, 2015.
Krause, Neal, and Elena Bastida. 2011. "Church-Based Social Relationships, Belonging, and Health among Older Mexican Americans." *Journal for the Scientific Study of Religion* 50(2): 397–409. doi:10.1007/s13644-011-0008-3.
Lambertville Beacon. 1999. "Shad Fest '99," April. Insert.
Landoll, Ryan R., Annette M. La Greca, and Betty S. Lai. 2013. "Aversive Peer Experiences on Social Networking Sites: Development of the Social Networking–Peer Experiences Questionnaire (SN-PEQ)." *Journal of Research on Adolescence* 23(4): 695–705. doi:10.1111/jora.12022.
Lattanzi Shutika, Debra. 2011. *Beyond the Borderlands: Migration and Belonging in the United States and Mexico.* Berkeley: University of California Press.

Laursen, Erik K., and Sasha Yazdgerdi. 2012. "Autism and Belonging." *Reclaiming Children and Youth* 21(2): 44–47. Academic Search Premier (79318060).

Lawless, Elaine J. 1993. *Holy Women, Wholly Women: Sharing Ministries through Life Stories and Reciprocal Ethnography.* Philadelphia: University of Pennsylvania Press.

Levine, Peter. 2010. "Teaching and Learning Civility." *New Directions for Higher Education* 152: 11–17. doi:10.1002/he.407.

Mack, Edward J. 1989. "Communications." In *The First 275 Years of Hunterdon County, 1714–1989*, 10–12. Flemington, NJ: Hunterdon County Cultural and Heritage Commission.

Maines, David R., and Jeffrey C. Bridger. 1992. "Narratives, Community, and Land Use Decisions." *Social Science Journal* 29(4): 363–80. Academic Search Premier (9305055050).

Mastrich, James, Yvonne Warren, and George Kline. 1996. *Images of America: Lambertville and New Hope.* Dover, NH: Arcadia.

Mayfield, Les, dir. 1994. *Miracle on 34th Street.* Twentieth Century Fox.

McPhee, John. 2002. *The Founding Fish.* New York: Farrar, Straus and Giroux.

Murphy, Sean Patrick. 2007. "Come Out and Enjoy Lambertville's Shad Festival." *Trentonian*, April 28. At http://www.trentonian.com/article/TT/20070428/TMP02/304289994.

Nanda, Serena, and Richard Warms. 2010. *Cultural Anthropology.* 10th ed. Boston: Wadsworth.

National Oceanic and Atmospheric Administration (NOAA). 2015. "Final Report: Flood Risk and Uncertainty; Assessing National Weather Service Flood Forecast and Warning Tools." Prepared by the Nurture Nature Center and RMC Research Corporation. Silver Spring, MD: National Oceanic and Atmospheric Administration.

Nielsen, Joyce McCarl. 1990. Introduction to *Feminist Research Methods: Exemplary Readings in the Social Sciences.* Boulder, CO: Westview Press.

Oore, Debra Gilin, et al. 2010. "When Respect Deteriorates: Incivility as a Moderator of the Stressor-Strain Relationship among Hospital Workers." *Journal of Nursing Management* 18(8): 878–88. doi:10.1111/j.1365-2834.2010.01139.x.

Ostrove, Joan M., Abigail J. Stewart, and Nicola L. Curtin. 2011. "Social Class and Belonging: Implications for Graduate Students' Career Aspirations." *Journal of Higher Education* 82(6) (November–December): 748–74. Academic Search Premier (67123530).

Ottenhoff, Patrick. 2017. "Week 52: Wakefield Ruritan Club; 69th Annual Shad Planking." 52 Week Season. At http://www.52weekseason.com/home/2017/4/14/shad-planking.

Oxford English Dictionary. (1971) 1982. Compact ed., s.v. "anomie."

Oxford English Dictionary. (1971) 1982. Compact ed., s.v. "touchstone."

Parish, Stan. 2011. "Traditional Tacos, and More." *New York Times*, March 18. Academic Search Premier (59395716).

Petrie, Alfred G. (1949) 1970. *Lambertville, New Jersey: From the Beginning as Coryell's Ferry.* Lambertville, NJ: n.p.

Pfeiffer, C. Boyd. 1975. *Shad Fishing.* New York: Crown.

Porath, Christine, Debbie MacInnis, and Valerie Folkes. 2010. "Witnessing Incivility among Employees: Effects on Consumer Anger and Negative Inferences about Companies." *Journal of Consumer Research* 37(2): 292–303. Academic Search Premier (52888684).

Pratt, Hawley, dir. 1972. *The Lorax.* Written by Dr. Seuss. Columbia Broadcasting System.

Prezza, Miretta, Matilde Amici, Tiziana Roberti, and Gloria Tedeschi. 2001. "Sense of Community Referred to the Whole Town: Its Relations with Neighboring, Loneliness, Life Satisfaction, and Area of Residence." *Journal of Community Psychology* 29(1): 29–52. doi:10.1002/1520-6629(200101)29:1<29::AID-JCOP3>3.0.CO;2-C.

Radner, Joan Newlon, and Susan Lanser. 1993. "Strategies of Coding in Women's Cultures." In *Feminist Messages: Coding in Women's Folk Culture*, edited by Joan Newlon Radner, 1–29. Urbana: University of Illinois Press.

Renaud, Chris, and Kyle Balda, dirs. 2012. *The Lorax*. Written by Dr. Seuss and Cinco Paul. Universal Pictures.

Richman, Laura Smart, Michelle vanDellen, and Wendy Wood. 2011. "How Women Cope: Being a Numerical Minority in a Male-Dominated Profession." *Journal of Social Issues* 67(3): 492–509. doi:10.1111/j.1540-4560.2011.01711.x.

Roberts, James A., and Meredith E. David. 2016. "My Life Has Become a Major Distraction from My Cell Phone: Partner Phubbing and Relationship Satisfaction among Romantic Partners." *Computers in Human Behavior* 54: 134–41. doi:10.1016/j.chb.2015.07.058.

Robins, Douglas M., Clinton R. Sanders, and Spencer E. Cahill. 1991. "Dogs and Their People: Pet-Facilitated Interaction in a Public Setting." *Journal of Contemporary Ethnography* 20(3): 3–25. doi:10.1177/089124191020001001.

Robinson, Mairi, ed. 1985. *The Concise Scots Dictionary*, s.v. "Crack, Crak &c, craik &c," 120. Aberdeen, Scotland: Aberdeen University Press.

Rodríguez, Sylvia. 1998. "Festival Time and Plaza Space." *Journal of American Folklore* 111(439): 39–57.

Rosenberg, Neil V. 1996. "Strategy and Tactics in Fieldwork: The Whole Don Messer Show." In *The World Observed: Reflections on the Fieldwork Process*, edited by Bruce Jackson and Edward D. Ives, 144–58. Urbana: University of Illinois Press.

Rosenthal, Lisa, Sheri Levy, Bonita London, Marci Lobel, and Cartney Bazile. 2013. "In Pursuit of the MD: The Impact of Role Models, Identity Compatibility, and Belonging among Undergraduate Women." *Sex Roles* 68(7–8): 464–73. doi:10.1007/s11199-012-0257-9.

Ryden, Kent C. 1993. *Mapping the Invisible Landscape: Folklore, Writing, and Sense of Place*. Iowa City: University of Iowa Press.

Savage, Mike, Gaynor Bagnall, and Brian Longhurst. 2005. *Globalization and Belonging*. London: Sage.

Scott, James C. 1990. *Domination and the Arts of Resistance: Hidden Transcripts*. New Haven, CT: Yale University Press.

Seaton, George, dir. 1947. *Miracle on 34th Street*. Twentieth Century Fox.

Selberg, Torunn. 2006. "Festivals as Celebrations of Place in Modern Society: Two Examples from Norway." *Folklore* 117(3): 297–312. At http://www.jstor.org/stable/30035376.

Shafer, Mary A. 2005. *Devastation on the Delaware: Stories and Images of the Deadline Flood of 1955*. Riegelsville, PA: Word Forge Books.

Smithsonian Institution. 2004. "Water Ways: Mid-Atlantic Maritime Communities." Smithsonian Folklife Festival. At http://www.festival.si.edu/2004/water-ways/smithsonian. Accessed September 26, 2015.

Spain, Daphne. 1993. "Been-Heres versus Come-Heres: Negotiating Conflicting Community Identities." *Journal of the American Planning Association* 59(2): 156–71. Academic Search Premier (9608226862).

Stone, Kay F. 1998. *Burning Brightly: New Light on Old Tales Told Today*. Toronto: Broadview.

Stutz, Bruce. 1992. *Natural Lives, Modern Times: People and Places of the Delaware River*. New York: Crown.

Swan Creek Rowing Club. 2015. "Frequently Asked Questions." At http://www.swancreek rowing.com/frequently-asked-questions/.

Tannen, Deborah. 1990. *You Just Don't Understand: Men and Women in Conversation*. New York: Ballantine.

Thomée, Sara, Annika Härenstam, and Mats Hagberg. 2011. "Mobile Phone Use and Stress, Sleep Disturbances, and Symptoms of Depression among Young Adults: A Prospective Cohort Study." *BMC Public Health* 11(1): 66–76. doi:10.1186/1471-2458-11-66.

Today's Sunbeam. 2012. "Atlantic Sturgeon in the Delaware River to Be Listed an Endangered Species." *Today's Sunbeam*, January 31. At http://www.nj.com/salem/index.ssf/2012/01/atlantic_sturgeon_in_the_delaw.html.

Tompkins, Jane. 1986. "'Indians': Textualism, Morality, and the Problem of History." *Critical Inquiry* 13(1): 101–19. At http://www.jstor.org/stable/1343557.

Tuan, Yi-Fu. 1977. *Space and Place: The Perspective of Experience*. Minneapolis: University of Minnesota Press.

Turkle, Sherry. 2011. *Alone Together: Why We Expect More from Technology and Less from Each Other*. New York: Basic Books.

Turkle, Sherry. 2015. *Reclaiming Conversation: The Power of Talk in a Digital Age*. New York: Penguin Press.

Turner, Edith L. B., with William Blodgett, Singleton Kahona, and Fideli Benwa. 1992. *Experiencing Ritual: A New Interpretation of African Healing*. Philadelphia: University of Pennsylvania Press.

Turner, Victor. 1974. "Social Dramas and Ritual Metaphors." In *Drama Fields and Metaphors*, 23–59. Ithaca, NY: Cornell University Press.

Turner, Victor. (1969) 1995. *The Ritual Process: Structure and Anti-Structure*. Hawthorne, NY: Aldine De Gruyter.

US Census Bureau. 2017. "QuickFacts." At https://www.census.gov/quickfacts/. Accessed June 27, 2017.

Vozzella, Laura. 2016. "Shad Planking, a Venerable Va. Political Confab, Tries to Reel In a New Crowd." *Washington Post*, April 23, 2016.

Warren, Yvonne, and Lou Toboz. 1998. *Lambertville*. Somerville, NJ: Aesthetic.

Wiborg, Agnete. 2004. "Place, Nature, and Migration: Students' Attachment to Their Rural Home Places." *Sociologia Ruralis* 44(4): 416–32. doi:10.1111/j.1467-9523.2004.00284.x.

Zimmern, Andrew. 2015. "Philadelphia: Shad Cakes, Krak and Kishke." *Bizarre Foods with Andrew Zimmern*, Travel Channel. https://www.travelchannel.com/shows/bizarre-foods/episodes/philadelphia-shad-cakes-krak-and-kishke.

Newspapers Consulted

Selected articles and advertisements are taken from the following, as cited in the text:

Bucks County (PA) Courier News
Hunterdon (NJ) Democrat
Hunterdon (NJ) Observer
Lambertville (NJ) Beacon
New Hope (PA) Gazette
New York Times
Trenton Times

INDEX

Page numbers in *italics* refer to figures and tables.

A

anecdotes. *See* character anecdotes
anomie, 194–95, 197, 210–11
anti-structure and structure, 142–45, 226ch6n1, 226ch6n2
Archibald, Robert, 171, 195, 215
arts, 10–11, 12; culture as, 205; stories, 109
Asia, South: and customers, 26. *See also* ethnicity and race
audience, xxi, 31; and Big Stories, 32; children as, 106, 208; connecting with, 60, 83, 87, 107, 110, 128, 207; for haul, 134, 136; influence on storytelling, 89–90, 93, 111; mixed gender, 22; multiple, 37; role-switching with teller, 127–28, 146, 182; at Shad Fest, 11, 76; for touchstone story, 123–25

B

Bacorn, Brandon, 152–53
Bacorn, Ken, 152–53
Bagnall, Gaynor, 170, 173–75, 189. *See also* elective belonging
Baker, Andrew, 15, 134–35, *219*
Baker, John, 15, *219*
Baker, Pam (Meserve), 15–18, 32, 42, 93–96, 153–54, *219*; and children, 85, 91, 181; and gender roles, 131, 163–64; and inclusivity, 121, 159, 164–65, 167–68; and media, 35, 59, 74; sense of place, 178, 189–91; on shad as trickster, 142–43; on sharing the island, 50, 52–54, 184, 190; and spirituality, 17, 67–68, 76; stewardship, 60–61, 153, 190, 202; storytelling role, 32, 112–16, 122, 128, 131–32, 152; storytelling style, 44, 101, 121–22, 128; and sustainability, 202, 206–7
Baker, Sarah (SaraBeth), 15, *219*; on connection, 107, 109; everyday storying with Keziah Groth-Tuft, 110; on kindness, 187; recruiting crew members, 151–52; role in storytelling, 106; on technology, 213–14; telling the Big Story of tradition, 152
Barr, Sue. *See* Meserve, Sue
Basso, Keith, 187–88
Bauman, Richard, 82, 107, 108, 128
Beacon. See *Lambertville Beacon*
Belanus, Betty, 31
belonging, 146–68, 173–76, 184–85, 188–90
Berg, Kathy, garden of, 6, 53–54, 134, 158, 180–81, *217*
Big Stories, 29–56. *See also individual names*
boats, 4, 23–24, 27, 59, 98; during flooding, 100; storytelling in, 132–35; vandalism of, 95–96
bridges: communities, 7–8; from Lambertville to New Hope, 55, 57, 91–92, 100, 101, 181–82, 184, 206, *217*. *See also* Lewis Island: footbridge to
Bruhn, John G., 99, 112, 149, 174

C

cabins, 6, 180, 212, *217*; in, 46, 138–39; privacy, 156–58; under the, xiv. *See also* sales

calmness, 67, 96, 143, 178–79. *See also* peacefulness
capitalism, 180, 196, 201, 203, 209–10
Cap'n Bill. *See* Lewis, William (Bill), Sr.
captains: and Big Story of the environment, 60–62, 63–64; character anecdotes about, 57–80; as figures, 57, 59, 60, 70–72, 156; intelligence of, 62–65; and media, 33, 59, 130–31; names of, 4; role during haul, 21, 24, 25, 46, 57; role of, xvii, 16, 30, 58–59, 65, 77, 84, 140, 180–82, 208; and shares, 27; succession of, 16, 35, 41, 60, 65, 70–71, 73, 76–77; as symbols, 58, 76; yelling, 96–98. *See also individual names*
Casey, Edward: anomie, 195, 215; implacement, 181, 183, 197–98, 209–10, 227ch8n2, 227ch9n4; thickening, 5, 159, 176, 179–80, 186, 198; time vs. space, 171–72
Cashman, Ray: character anecdotes as type, 57–59, 72; flat or round characters, 69–70, 125; on narrative and social processes, 76, 79–80, 142; outwitting outsiders story type, 63; purpose of narrative, 126; relationships with informants, xvi, 95; on stories and conversation, 22, 82, 119, 128, 142
Centenary Methodist Church, 17, 53, 67, 74–75, 204
character anecdotes, 57–80, 119; and community building, 79–80
characters: actors in story, 179; as moral trait, 59–60; round vs. flat, 69–70, 71; as type of person, 59
children and youth: behavioral expectations of, 42, 45, 157, 187, 208; benefit of peacefulness to, 178; as crew members, 20, 27, 42, 91, 152, 160–63; fishery kids, 42, 60, 154; and flood clean-up, 104, 169; interaction with adult crew members, 46, 59, 147, 150–51, 175–76; and Mrs. Berg's garden, 180–81; participation in storytelling, 83, 85, 91, 106, 110, 116, 133; participation in the haul, 24, 131–34, 140; taking care of, 49, 54, 102, 143, 175; and vandalism, 95–96; as visitors, 26, 50, 54, 96, 97, 123, 138, 155–56, 158, 181, 204
chitchat, 26, 109, 204
Christianity, 17, 53, 67–68, 83, 144–45, 161
civility, 226n5
—and behavioral standards, 41–56, 121–22
—and belonging, 149–50, 153–54, 158
—Big Story of, 29, 34, 43–56, 64, 79, 94–95, 121, 128, 157; and inclusivity, 161, 164–65, 167; and sense of place, 215
—and embeddedness, 206–8
—environmentalism as, 103–4
—and sense of place, 187
—and social interaction, 111, 168
—surveillance, 94–98
class, social, 9, 62–63, 160, 189. *See also* gentrification
cohabitancy, 176, 179
collaboration: within community, xv, 10, 54, 94, 110; with researchers, xiii, 38, 63; in storytelling, 83, 127–46. *See also* teamwork
commitment, xiii, xxi, 48–49, 147, 149–53, 209; to physical presence, 166, 174, 215–16; and Steve Meserve, 76–77; sweat equity as, 67, 190. *See also* persistence; stubbornness
community, xxi, 4, 80, 168, 174–76, 210
—and belonging, 146, 188
—Big Story of, 29, 32, 34–35, 43–56, 80, 94
—bridge, 7–8
—connections, 35, 105, 121, 149
—and cultural sustainability, 197–98, 201, 205–8, 210–11
—discourse, 107, 128
—and flooding, 99, 105
—reminiscence and, 119
—stories about, 4
—strengthening through narrative, xix, 14, 80, 82–83, 146, 174, 198, 216
—as support, 112, 149

conservation of island, 7. *See also* sustainability
contract: to collect data, 20–21, 38, 203–4; and fishery economics, 27, 207
Coryell Street, house on, 41, 100–101, 105, 116, 141, *217*
crews: collective identity of, 116; diversity of, 160–63, 200; membership, 41–42, 46, 147, 149–54, 158; paid vs. volunteer, 27, 46, 79, 136, 201, 203, 207; representing fishery, 46, 150–52, 155; retired, 27, 115–16, 119, 157, 208; roles in a haul, 114–15; size of, 27. *See also individual names*
cultural change, xx, xxi, 14, 161–63, 194, 199–201, 215. *See also* shift and drift
customers, 204; and belonging, 157–58; diversity of, 26–27, 164–68, 204–5; stories about, 94, 140. *See also* sales

D

data collection: and environmental concerns, 20–21, 36–38, 62–64, 203–4, *221–24*; and narrative, 26, 125, 215; as process, 20–21, 78, 138. *See also* record-keeping
Dégh, Linda, ix, 29–30
Delaware River, 3–5, 35, 39–40, 69, 201; attachment to, 77, 165; bridge communities, 7; and environmental concerns, 11, 37–39; and industrialization, 8–9; Little Gap, 7; Scudder's Falls, 7, 181; as site of shad fishing, 18, 35, 39, 93, 181, 201; and spirituality, 54–55; Trenton Falls, 7; and Washington's crossing, 8. *See also* flooding; river
Delaware River Basin Commission, 13, 103
Delaware River Basin Fish and Wildlife Management Cooperative, 20. *See also* contract
diversity, 26–27, 94, 159–68, 174, 204–8. *See also* ethnicity and race; inclusivity
dogs, 111–22, 131, 133, 151
dogwalkers, 52, 54, 95, 158, 174–75, 208
Dunaj, Greg, 154

E

economics: and belonging, 148; and diversity, 160, 161; in Lambertville, 9–10, 13–14; and Lewis Fishery, 19–21, 27–28, 38, 51, 52, 91; and myth of progress, 195; and sustainability, 199–212
education: and civility, 96, 226n5; and diversity, 160, 163; as Lewis family value, 16–18, 65–66, 76, 160; and Lewis Fishery, 13, 31–32, 35, 36, 73, 202; and teacher's role in storytelling, 30
elective belonging, 173–74, 177, 188–90, 205, 215
embeddedness: civility as value of, 49; and narrative stewardship, xxi, 82, 168, 197–98, 206–9, 212, 216; processional storytelling, 127–46
emplacement, 227ch8n2. *See also* implacement
enactment, 30, 31–32. *See also* performance
enculturation, 5, 180, 186
environment, 175
—and belonging, 148–49
—Big Story of, 27, 29, 32, 34–40, 45, 56, 84, 88, 144, 146, 148, 168, 202; stories about captains and, 60–65, 70, 75–76
—and microlegends, 84–88, 103–4
—and narrative stewardship system, xxi, 4
—organizations, 13, 31
—and sense of place, 178, 186, 189
—and Shad Fest, 11, 13
—*See also* sustainability
ethics, xvi–xvii, xxi, 64, 107, 176, 183, 209. *See also* civility: Big Story of
ethnicity and race, xv; African Americans, 26, 164; Americans, 201; Asians, 204–5; Chinese, 26; English, 23; European Americans, xv, 26; and foodways, 26–27; Hispanics, 13–14, 26–27, 204; Indians (Asian), 26, 164–67, 207; mainstream, xiii, xx, 198; Mexicans, 13–14; Swiss, 180. *See also* diversity; inclusivity

everyday storying, 81, 95, 107–26; and belonging, 157–58, 165, 168; and cultural sustainability, 198, 207, 214; and processional storytelling, 127, 139–40, 144; and sense of place, 189. *See also* reminiscence; touchstone stories; troubles

F

faith, xxi, 63–64, 69, 79, 96, 160, 193. *See also* Christianity; commitment; religion; spirituality

feminism: and ethnography, xvi; scholarship, 199–200, 207. *See also* gender; women

Finkle, Frances, 164

Finkle, Joseph, 164

fish: distribution of, 19, 26, 47–49, 138–39, 157–58, 165–66; first, 43; food vs. sport (game), 25, 85; species sold, 25

fisheries, other, 40; Green Bank Island, 15; Liberty, 15; Malta, 19, 65; Point, 15

fishing season: basics, 18–20, 33, 64, 68, 88, 100, 201; and belonging, 149–50, 164; and Big Stories, 42–43; and everyday storying, 106, 108, 112, 114, 116; overlap with other events, 10, 13, 31, 160–61; and processional storytelling, 127, 135–39, 141, 142, 144, 146; and Steve Meserve, 76, 77, 78. *See also* springtime: Big Story of

flooding: and closing island, 52–53, 183, 184, 190; and erosion, 44, 54–55, 177, 180, 185, 227ch8n1; and microlegends, 90, 94, 99–105, 144, 198; record floods, 68, 100–101, 105, 169–70; and sense of place, 3, 5, 6, 64, 144, 169–70, 177–78, 180, 185; and sustainability, 211–12. *See also* Lambertville–New Hope bridge; Lewis Island

folkloristics, xiii, xv–xvi, xix, 14, 29, 170–71, 196, 205

foodways, 26–27, 139, 164–66, 168, 205, 207; shad-planking, 26, 124–25

Founding Fish, The, 39, 84

French scenes, 129–30, 133

G

Garber, Daniel, 50

Garczynski, Sue (Lewis), 77, *219*; breaking gender barriers, 15, 71, 161–62; co-owner of island and fishery, 5, 16

Geertz, Clifford, xxi, 5, 30, 193, 197–98, 225ch6n4

gender, 148, 167, 207; and communication, 112, 128–29, 135, 137–38; inclusion, (in)equity, xv, 59, 63, 71, 74, 147, 161–64, 189, 200, 204; roles in storytelling, 21, 112–14. *See also* masculinity; women

Genthner, Tim, 152; and everyday storying, 110–11, 116; and microlegends, 81, 91–94

gentrification, 10, 13–14, 160

girls, role in count, 25, 65, 67, 131, 161–62, 168. *See also* children and youth; women

Glassie, Henry, xxi, 22, 58, 63, 193, 194

Goffman, Erving, 121, 168

Grammy. *See* Lewis, Nell (Eleanor Windle)

Grampy. *See* Lewis, Fred

Groth-Tuft, Adelaide: as crew member, 130; and flooding, 104; "The Night We Caught a Fish as Big as Adelaide," 85; and photography, xvii; roles in storytelling, 83, 106, 132–33

Groth-Tuft, Keziah: and Arabic, 167; on behavioral expectations, 175–76, 187, 190, 208; everyday storying, 110; and flooding, 105; function of storytelling, 145; observations of, 45, 46, 65, 84, 91, 111, 120, 122, 139, 157–58; and roles in storytelling, 83, 116, 128, 132–33, 141; on sense of place, 213–14; and Shad Fest, 155–57

H

Hancock, New York, 18, 35, 93

Hartpence, Norman (Sparky), 115

haul: basic process, 21–26; and processional storytelling, 127–46; as ritual, 75, 131, 136, 144–45

haul seine fishing method, 4, 13, 15, 18–26; awareness of, 98, 147; and Big Stories, 40, 42, 71–72; and place, 5, 177, 180–81; and sustainability, 201, 203–4, 206, 209
Hawkes, Jonathan, 198–206
Hazen, Joseph (Joe), 33–34, 71–72
Hazen family, 33–34
hierarchy, 58–59, 74, 78, 124, 162. *See also* gender
Holcombe Island, 5, 15, 39, 187–88
home: Lewis Island as, 5, 24, 173; and research, xiv, xviii; and sense of place, 169–72, 175, 181–82, 189, 209
Hufford, Mary, 125, 226ch6n1, 226ch6n4; on building collective history, 142; stories within conversation, 109, 128; on Story-realm, 115–16, 159
humanities, xxi, 195–96, 199, 210
humanity, 50, 55–56, 57, 94, 119, 193–94
humility, 80, 98, 196
humor: and Fred, 66–67, 68–69, 84; joke about shad-planking, 124–25; joking and teasing, 78–79, 83, 97–98, 116–19, 121, 137, 150–51, 162; and wonder stories, 89–90, 102
Hunterdon County, New Jersey, 18, 32, 204, 226ch4n1
Hunterdon Democrat, 33, 38, 59

I

identity, xx; and belonging, 148, 168; as captain, 71, 73, 77, 114; and elective belonging, 174, 188–91, 205, 215; Lewis family, 52–53; and sense of place, 53, 171–74, 188–91
implacement, 172, 181, 183, 186, 190, 197–98, 210–11, 213, 215, 227ch8n2
impression management, 85, 168, 198
inclusion, 26, 110, 128, 146–47, 149. *See also* diversity; gender
Indians (American). *See* Native Americans
Indians (Asian). *See* ethnicity and race
integrity, 16–18, 66–67, 83–84, 160–61
invisibility, xxi, 197–98, 204–9

invisible landscape, xviii, xx, 5, 53, 124, 145, 159
Isler, John (Johnny), 27, 86, 90, 109–10, 115–16, 152
Iuvone, Mary, 32–33, 40, 42, 154–55, 177, 183–84, 189

J

Jews, 125, 161, 164–65, 204
Johnston, Robert, 157
Johnstone, Barbara: mainstream culture, xiii; myth, 55; narrative and community, 174–75, 188; narrative core, 82; narrative tension, 215; personal experience narratives, 106; public vs. private stories, 146, 173; and studying one's own culture, xvi. *See also* texture, narrative
joking. *See* humor

K

kindness: and belonging, 149, 160–61, 164; and Big Stories, 45–47, 49, 55–56, 209; as Lewis trait, 17, 60; and sense of place, 117, 169, 175–76, 183, 187. *See also* civility
Kiriluk-Hill, Renée: covering fishing season, 33–35, 84; environment, 37–38; history, 40, 41, 175; Lewis family, 16–17, 59, 65, 73–74, 187; photography, 162, 183
Kroemmelbein, Diane: childhood experiences, 154; memories of Isler, 86, 109–10, 115–16; role in storytelling, 87; sense of place, 177; Shad Fest crowds, 155–56
Kroemmelbein, Ted, 38, 117, 137, 160, 203; crewmember role, 25–26, 92, 150–52, 154, 174, 189; and everyday storying, 110–11; memories of Isler, 115; microlegends, 86–87, 89, 92–93, 96–97; on representing fishery, 46, 86, 150–52; sense of place, 177–78, 182–83; Shad Fest crowds, 155–56, 184; and spring, 135–36, 143

L

Lambertville, New Jersey, 5–10, 17, 76, 160, 181, 189, 190; community values, 35,

159–60, 175; and diversity, 13–14, 164, 204–5; flooding in, 104, 170; gentrification of, 10, 13–14; as tourist destination, 157. *See also* Shad Fest
Lambertville Beacon, 21, 32, 33–34, 71–72, 109
Lambertville–New Hope bridge, 55, 57, 91–92, 100, 101, 181–82, 184, 206, *217*
language, use of: "bushwhacker," 21; "coming down," 114; "drifter," 159; "giffen," 47; "like," 107; "true fishing dog," 120; types of holes in net, 22; world languages on Lewis Island, 165; "you know," xvi
Lattanzi Shutika, Debra, 10, 14, 147–49, 168, 173, 206
legends, 4, 29–30, 80, 87; and character anecdotes, 57–58, 63–64, 70–73. *See also* microlegends
Lewis, Cliff (Fred Clifford), 15, 16, 71, 113, *219*, 227ch8n1
Lewis, Donald, 15, 101
Lewis, Edith (Eek), 15, 118, 186, *219*
Lewis, Eleanor (Windle). *See* Lewis, Nell
Lewis, Fred, 11–12, 15–22, 38–39, 40, 48, 63, 84, 87–89, 94, 95–96, 101, 161, 201, 202, *219*; and belonging, 162, 164; and Bill Jr., 5–6, 19, 41, 64; character anecdotes about, 65–69; death of, 52–53, 72–76; everyday storying, 109–10, 114–16, 122–23; and faith, 54, 67–68, 76, 96; as island resident, 6, 52–53; leadership role of, xiv, 4, 16, 65, 77; and microlegends, 90–91, 99; and Nell, 6, 17, 59; as public figure, 12, 31, 33, 59, 62, 70–76; and sense of place, 179, 180, 182, 184; and standards of behavior, 41, 44–45, 114, 151; as story character, 55, 179; and Union Paper Mill, 9–10, 19
Lewis, Muriel. *See* Meserve, Muriel (Lewis)
Lewis, Nell (Eleanor Windle), xiv, 15, 18, 31, 41, 51, 60–66, 88, 99, 129, 191, *219*; and faith, 17, 67, 152; and Fred, 6, 17, 52, 59, 74; and gender barriers, 161; island resident,

52–53, 101, 184; as owner, 16, 102, 141, 180; and sales, 48, 118–19, 141; and storytelling, 112–14, 116, 122, 139
Lewis, Sue. *See* Garczynski, Sue (Lewis)
Lewis, Theodore, 6, 15, 42, 60, 62, 65, 119, 180
Lewis, William (Bill), Jr., 5–6, 15, 19, 41, 64, *219*
Lewis, William (Bill), Sr., 22, 70–72, 77, 80, 115, *219*; and Big Story of the environment, 4, 36–39, 61–63, 86–88, 93, 99, 201–2, 226ch3n1; as carpenter, 23, 59, 64–65; character anecdotes about, 58–65; crew of, 18–19, 42, 43, 152; as fishery founder and owner, 4, 15, 33, 39, 41, 49, 161; and record-keeping, 21, 36–37, 63; and standards of behavior, 41, 45, 61. *See also* mandate to share the island
Lewis family, *219*; and belonging, 147, 149–50, 152, 189; and Big Stories, 29, 31, 34, 36, 39, 43, 96; collective identity of, 14–18, 23, 41, 158, 160, 205; everyday storying, 124, 128–29; and microlegends, 99, 101; as public figures, 60, 74, 112, 187. *See also individual names*
Lewis Island, *217*
—behavioral standards on, 43–45
—community activities on, 50–53, 185–86, 227ch8n1
—description of, 3–7
—erosion of, 43, 52–55, 180
—flooding, 99–105
—footbridge to, 3–4, 6, 22, 51–52, 155, 206, *217*; flooding and rebuilding of, 4, 6, 41, 43, 55, 64, 68–69, 99–105, 150, 169, 180, 211–12; gates of and closing, 52–53, 184–85, 190; threshold to, xiii, 3, 92, 117, 130, 140, 148, 158–59, 167, 169, 179, 185
—house on, 135, 187, 212, *217*; building of, 5–6, 64; and flooding, 52, 99–101, 104, 169–70, 190; landmark during haul, 6, 92, 130, 182; and privacy, 49, 158, 184
—and invisibility, 3, 206, 208

—mandate to share, 13, 26, 29, 43, 49–54, 64, 124, 147, 176, 184, 205–6 (*see also* Berg, Kathy: garden of)
—ownership of, 5, 16, 29, 39, 149
—representing, 150–52, 155
—and sense of place, 123–24, 144, 147, 169–92, 194
—and spirituality, xix, 53–56, 67, 179, 189, 227ch8n3
—vandalism on, 94–96
—*See also* Holcombe Island
license, fishing: anchored to place, 148, 181; date limitations, 18, 88; preserving, 20, 41, 61, 147, 201, 203; species limitations, 25
liminal space, 77, 144, 155, 178, 180
Longhurst, Brian, 170, 173–75, 189. *See also* elective belonging
Lorax, The, 193, 202

M

mainstream culture. *See* ethnicity and race
Marriott, Jack, 75, 116, 119, 152
Marriott, Ricky, 119, 152
masculinity, 137, 202; activities, 113, 138, 168, 189; and crew, 46, 49, 58–59, 97
material culture, 115, 179–82, 185–87, 205
McMichael, Kelly, 86, 135–36, 155–56, 184–85
media: and flooding, 94, 100, 103; and Lewis Fishery, 16, 31–35, 42, 59, 74, 80, 98, 130, 152, 154–55, 157, 207; and sense of place, 171, 183–84; and sexism, 163; and Shad Fest, 11, 13, 26, 31, 38–39, 152, 206; social, 109, 195, 213–14. *See also* technology
Meserve, David, 16, 71, *219*
Meserve, Muriel (Lewis), xviii–xix, 5, 11, 15–18, 25, 41, 45, 53, 60, 64–65, 80, 116, 153, 163, *219*, 226n7; and Big Stories, 62–64, 225ch2n1; birthday party, 52–53, 190–91; and everyday storying, 112–15, 118–19, 122; and microlegends, 84–85, 88, 95, 100, 140–41; sense of place, 56, 172–73, 178–79, 184–86, 227ch8n1; and sharing island, 49–51, 176; and spring, 42–43, 190; style as storyteller, 44, 128; and sustainability, 199, 202, 211; as teacher, 31–32, 35–36, 66–67; and women's roles, 161–62. *See also* Centenary Methodist Church
Meserve, Pam. *See* Baker, Pam (Meserve)
Meserve, Steve (Stephen), 15–18, 27, 41, 108, 133–35, 139, 188, 189, *219*, 225n3; and Big Stories, 35–36; character anecdotes about, 71, 75–79, 97–98; and data management, 78, 224; and everyday storying, 111, 114–17, 122; and his crew, 45, 152–53; and leadership role, 4, 15, 35, 41–42, 162; and microlegends, 89–91, 94; as public figure, 31, 33, 35, 70–73; sense of place, 178–80, 184; storytelling role of, 83, 113, 130–32, 140–41; and Sue, 59, 77; and sustainability, 202–3, 215–16; technology and tools and materials, 22–23, 212–13
Meserve, Sue (Barr), 15–18, 43, *219*; under the cabin, 48, 138–39, 166–68; as carpenter and technical director, 23, 46, 59, 110; and children, 26, 83, 132–33; and everyday storying, 110–12, 117, 122, 125; and gender equity, xv, 162–63; and microlegends, 89, 94; observations of, 23, 41, 108, 121, 133–35, 141, 188, 225n3; and sense of place, 180, 185–86, 227ch8n1; and Steve, 59, 77; technology, 211–13
Mexicans. *See* ethnicity and race
microlegends, 78, 81–106; and belonging, 159, 168, 198; and everyday storying, 109, 119; and processional storytelling, 127, 144. *See also* surveillance stories; wonder stories
Muslims, 167–68
myths, 13; and Big Stories, 29–30, 39, 55–56; and captain stories, 57, 58, 68, 71–73; of progress, xx, 195, 198–99, 201, 209–10; and ritual, 30, 129, 215

N

narrative stewardship, xxi
narrative typology, xix, xxi, 29–30, 57–58
Native Americans, 5, 21, 40, 50, 53, 72
New Hope, Pennsylvania: fisheries, 15–16, 23; former home of Lewis family, 15, 22; and Lambertville, 7–10, 51, 66. *See also* arts; Union Paper Mill
New Jersey, 10, 26; Department of Environmental Protection, 38, 203; Division of Fish and Wildlife, 5, 38; fisheries, 18–19
New York Times, 14, 32, 59
nostalgia, 195, 197, 212

P

peacefulness, 169, 170, 173, 176, 178–79, 183, 187, 189. *See also* calmness
performance, 35, 44, 67, 108, 117, 125, 127, 156–57. *See also* enactment
persistence, 4, 37, 60–62, 64, 144. *See also* stubbornness
personal experience narratives (PENs), 30, 55, 58, 81, 106, 108–10. *See also* touchstone stories
photography, 32–33, 154–55, 162–63, 183–84
Poppy. *See* Lewis, William (Bill), Sr.
postmodernity, xvi, 15, 197, 199, 208, 214
preservation. *See* sustainability
private vs. public: space, 5, 49–50, 52, 122, 145, 149, 157–58, 184–86, 190; storytelling, 95, 134–35, 146, 173
problems. *See* troubles
processional storytelling, 116, 127–46, 190, 198

R

race. *See* ethnicity and race
record-keeping, 20–21, 26, 36–38, 62–63, 78, 85, 138, 221–24. *See also* data collection
religion: and diversity, 52, 160–61, 163, 164, 167, 204–5; experiences, 53–54, 179; processions, 144–45; services, 50, 208. *See also* Centenary Methodist Church; Christianity; spirituality
reminiscence, 114–16, 119, 122–23
reporters: covering Lewis Island, 33–36, 42, 58–59, 72–73, 78, 130, 138; and impact on storytelling, 155; and microlegends, 106; as storytellers, 30. *See also* Kiriluk-Hill, Renée
rites: birthday gatherings, 190; fishing as, 35; of passage, 42, 75, 78, 152, 176, 226ch6n3; of spring, 26, 33, 42–43, 54, 189
rituals, 32, 74, 86, 104, 189–90; and Big Stories, 35, 42–43; final report as, 140–42; involving fish, 26, 43, 164; as key concept, xiii, xix, xxi; and myth, 30, 128–29; and narrative, 31, 33, 44, 108, 115; and processional storytelling, 127–46; and touchstone stories, 122–24; visit to Lewis Island as, 42–43, 54, 215
rivers, 176; impact on storytelling, 122, 133–35; power and danger of, 23–24, 41, 43, 64, 87, 90, 99–105, 143–44, 169–70, 180, 211–12; as spiritual force, 53–55, 67–68, 74–75, 173, 179
Ryden, Kent, 145, 172, 188, 190; invisible landscape, xviii, xx, 5, 53, 123–24, 159, 186

S

sales, fish, 19–20, 26–27, 47–49, 138–39, 157–58, 165–66
Savage, Mike, 170, 173–75, 189. *See also* elective belonging
season, fishing. *See* fishing season
sense of place, 54, 121, 142, 169–91; and cultural sustainability, 194, 197–98, 204–5, 209–11, 215; and dogs, 121; and touchstone stories, 123–24. *See also* elective belonging
shad: American (*Alosa sapidissima*), 24–25; best-tasting, 181; gizzard, 25; similar fish

in South Asia, 26; as trickster figure, 142–43

Shad Fest, 9–13, 98, 125, 227; and Big Stories, 31, 34, 38–39; and death of Fred Lewis, 73–76; demonstration hauls, 13, 24, 32–33, 90; and environment, 11, 13, 38; founding of, 11–12; media coverage of, 26, 32–34, 38, 42, 71–72, 78, 130, 163; poster auction, 12; representing fishery at, 151–52; T-shirts, 11; visitors, 102, 154–57, 167, 184, 206

shad festivals, Bethlehem, 73, 125, 225n2

shad-planking, 26, 124–25, 225n2

shift and drift, 199–202, 211, 215–16

Shutika, Debra Lattanzi. *See* Lattanzi Shutika, Debra

Smithsonian Folklife Festival, 31, 102, 154, 205

spirituality, 43, 144–45, 186; and displacement, 99; as feature of Lewis Island, xix, 53–56, 67, 179, 189, 227ch8n3; as key concept, xix; numinous, sense of, 54, 67–68, 87; and river, 53–55, 67–68, 74–75, 173, 179. *See also* Centenary Methodist Church; Christianity; religion

springtime, Big Story of, 29, 34–36, 39, 42–43, 135–36. *See also* rites

standards of behavior, 50, 117, 151, 225n3; and captains, 41–42, 44–46, 61, 64–66. *See also* civility

storytellers: as role, 82, 106, 127; typology, 30

storytelling: collaboration in, 83, 127–29, 135–43, 146; as social process, 124, 127–29

structure and anti-structure, 142–45, 226ch6n1, 226ch6n2

stubbornness, 35–36, 41, 209, 216, 225ch2n1; character anecdotes about, 60–62, 64–66, 68–70, 72, 78, 80. *See also* persistence

surveillance stories, 44, 94–106, 140–41, 157–58; about flooding, 99–105; about vandalism, 94–99

sustainability, 193–216; cultural, xxi, 203–16; economic, 19–21, 199–212; environmental, xx, 201–3; social equity, 201, 204–5. *See also* environment

T

Tannen, Deborah, 112, 128, 137

teasing. *See* humor

technology: change at fishery, 23, 180, 211, 216; critical thinking about, 209–14; progress as myth, 195, 199, 201, 209, 211–12; and sense of place, 171, 174, 180, 213–15; widespread change, xx, 194–97. *See also* media

texture, narrative, 159, 175, 179, 186, 190, 198

thickening, 5, 159, 176, 179–80, 186, 190, 198

Thoreau, Henry David, 176

touchstone stories, 122–24, 130, 144–45, 146, 165, 168, 172, 176, 186, 189

tradition, Big Story of, 29, 34–36, 39–42, 73–74, 144, 152

traditional knowledge, 102, 114, 145, 182

Trenton, New Jersey, 7, 10, 18, 21, 93, 164, 226ch3n1

Trenton Times, 32, 74

Trentonian, 74

troubles: as problem-solving, 53; talk, 111–12, 134–35; walking out problems, 54, 179

Tuan, Yi-Fu, xx, 172, 177, 197

Tuft, Dan: everyday storying, 114, 116, 132; microlegends, 84, 92, 102; on privacy, 134, 185; representing fishery, 151–52, 154; sense of place, 178; and technology use, 212–13

Turkle, Sherry, 196–97, 210

Turner, Edith, xviii

Turner, Victor, 54

U

Uncle Dory. *See* Lewis, Theodore

Union Paper Mill, 9, 16, 19, 42, 64, 152; everyday storying about, 109–10; and first fish, 43

V

visitors, 26, 112–13; common misconceptions of, 11, 19; diversity of, 26; fishermen as, 25, 111, 137; participation in narrative stewardship, 31–32, 128, 131, 158; role in processional storytelling, 137–38; and sense of place, 173; during Shad Fest, 31, 154–57, 184. *See also* touchstone stories

W

Wiborg, Agnete, 171–72, 174, 189
Williamson, Sis, 118–19
women: roles at fishery, xiv, 16, 74, 118–19; roles in storytelling, 112–14; wives of captains, 59. *See also* fish: distribution of; gender; sales
wonder stories, 83–94, 105; about calamities and mishaps, 90–94; and humor, 89–90; about unusual catches, 84–90
worldview, 99, 104, 171, 195, 197, 199, 201, 209

Y

youth. *See* children and youth

Z

Zimmern, Andrew, *Bizarre Foods with Andrew Zimmern*, 97–98, 205

Printed in the United States
By Bookmasters